THE SEMIMETALLIC

MONSTERLING

THAT CHANGED

OUR WORLD

The Semimetallic Monsterling that changed our World

and ... his journey since 1947 until 2050

English Edition B/W

Ettore Accenti

http://ettoreaccenti.blogspot.ch/

This book is available in the following formats:
- Paper book, black and white
- Paper book, color
- eBook Kindle

ETTORE ACCENTI EDITION

Ettore Accenti

The Semimetallic Monsterling that changed our World

English edition B/W

ISBN – 13: 978-1548986650   ISBN-10: 1548986658

Dedication and important author's note

This book is dedicated to my wife Eva who, with extreme patience, helped me to correct the English language. We decided to write this edition in English by combining our skills: my knowledge of technical lexicon with her knowledge of English literature ... hoping to have created a work in the language that we used so many times in our trips to US. We ask the reader an understanding for our effort.

The author

Since school age I became interested in electronic technology, beginning with the Transistor and then with everything derived from this electronic component as Integrated Circuits, Microprocessor and Personal Computers.

While I was a high school student and then a college student I published more than 50 articles. In 1969 I published my first book "From Transistors to Integrated Circuits", still on sale today for students of technical schools.

While attending the last year of the master on engineering at the Politecnico di Milano, I founded my first start up and once graduated I begun to represent several companies such as Intel, Apple Computer, AMD, HP, IBM, Intel, Philips, RCA, National Semiconductor and many others.

I participated for 30 years in the development of the electronic market and I had the opportunity to meet and collaborate with a large number of people in the Silicon Valley.

I have had the pleasure of knowing many interesting people, many of whom are now part of the Valley's history.

I remember:

At **Intel** , Robert Noyce, Gordon Moore, Federico Faggin, Stan Mazor, Tom Lawrence, Andy Grove, Ed Gelbach, Jack Carsten, Hank O'Hara, Jens Paulsen, Mike Markkula, Hal Feeney, Robert

Graham, George Adams, Ron Kerner, Jeff Miller, Robert McIntosh, Jim, Dodson, Rob Walker, George Schnear, Bob Derby,William Davidow, Jens Paulsen, Dennis Lundien, Ben Franklin, Jim Oliphant, Guy Debruine, David House, Maria Ligeti, Dave Williams, Jim Lalli, Dick Clover, Phil Spiegel, Bob Sumbs, Gerr Griese, Gary Webb, Gary Andersen, Lou Calcagno, Bernard Giroud, Jean Cloud Rivet, Tony Livingstone, Gary Webb.

At **Apple Computer:** Mike Markkula, Steve Jobs, Mike Spindler, William Broderick, Bob Dijkman, Richard Haas.

At **AMD:** Terry Jones, Ben Anixter, E, Brown

At: **Hewlett Packard** Jan Black, Al Hockley

….. and many more.

Thanks to the great experience gained in the technology and international markets, since 1994 I have been carrying out an activity of helping small and medium-size companies to gain presence all over the world.

Moreover, living in Lugano, in the beautiful corner of Switzerland, which is called Canton Ticino only half an hour from Italy, I love writing and publishing articles and books on topics that I like most.

All this is possible because my four children are now independent, my ten grandchildren are well cared for by their parents and my wife who knows languages can help me with everything a self publisher like me needs.

SUMMARY

About this Book

Anyone, a student, a professional, a technology lover will find in this book, unique in the world, to know more than seventy years of technology and to predict what will happen in this area until the year 2050.

The text is a condensate of historical facts, technical information and autobiographical stories, many of them unpublished.

As we will see, in the year 2050, the complexity of the microchips will cross the complexity of the human brain.

All this technological progress that begun and developed in the twenty century is due to the invention of the Transistor which in this book we nicknamed the "Semimetallic Monsterling".

The wonderful products that billions of people use every day are made with our "Monsterlings" and the purpose of this book is to reveal the mysteries of how everything was originated since his birth in the year 1947.

In this text, addressed to everybody, and, above all, addressed to young people, specialized terminology is avoided as far as possible for its broad comprehension without diminishing the validity of the text.

Historical, autobiographical and technical facts are described here in great detail, linked by both temporal and logical succession for a better understanding.

Some chapters are dedicated to readers wishing to understand how various devices such radio, Transistors, integrated circuits, microprocessors, personal computers work. The purpose is to reveal this magic world to both curious readers and enthusiastic technicians.

The history of electronic devices and how they have evolved over time are part of other chapters where the names of the genius men that have been the creators of all this progress are mentioned: William Shockley, Bob Noyce, Gordon Moore, Andy Grove, Steve Jobs are just a few.

At the end of the book, having visited so many times the Silicon Valley, I offer my point of view about that wonderful valley and my opinion why this California corner, cradle of large companies such as Intel, Apple Computer, Microsoft, Google, Facebook, is still so important for everything that concerns technology.

Ample space is dedicated to my many meetings with important scientists of the Silicon Valley such as Gordon Moore, Bob Noyce, Federico Faggin and many others.

I describe some curious and funny episodes, all widely documented with original images. At the end the reader will have a clear view of what is the effect of the Dr. Moore's law, of which so much is spoken today.

How the Idea of this Book came to my Mind

Everything began on the day when, speaking of various topics with my "number three" grandson, as I call him not to be confused with the names of the other nine grandchildren, he expressed a question that shocked me.

Lorenzo, ten years old, had been interested in some small technical issues for a long time and used to ask me difficult question despite his young age. That day he was playing an iPhone4s that his mom, having bought an iPhone5, had given him.

He considered his iPhone a precious jewel. Turning to me suddenly he asked me: "Grandpa, you always talk about the electronics of your age, please tell me what's inside that makes it work!"

"Gulp!" I thought , "but where do I start?" So I took his le iPhone on my hand, trying to find a curious answer for him and I exclaimed "Do you know that you have in your hand a few hundred millions electronic components?". With this words I hoped to rise his imagination.

I also thought that I would have to show him the following image that summarizes the progress of humanity and that projects it up to the year 2050, the same image that will be explained at the end of this book, but perhaps it was still too early for him.

Despite this concept was rather difficult for a boy of his age, I showed him the drawing trying to give him a simple explanation.

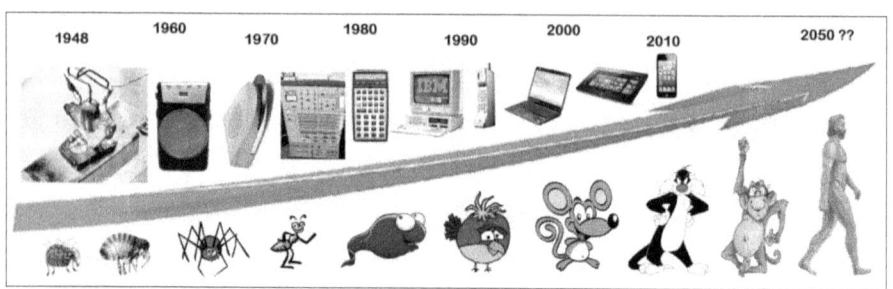

Image shown to my grandson. Scale of living beings, compared with the development of electronics (Moore's Law and the year 2050)

After my ten minutes talk, I hoped he would ask me the rest of the story, but I got this answer: "Grandpa, don't joke with me, I am not five years old, please speak seriously".

Double gulp! I had said a great story but how could I explain him that it was really true?

Also, if I had not told him something credible, this nice little grandson would take me for a visionary old man and this fame would circulate among all the other grandchildren with whom he was connected via sms, email, facebook and all those little devilries that today they can use with utter skill.

I could not just give up, I had to say something credible to a child of that age, but there was a double problem. If I had explained the concept in a way too simple, like a fairy tale, he would be offended because he considered himself an adult capable of understanding anything, if I had entered the technical details he would have told me: "Granpa, if you don't know this subject, don't try to cheat me, talking in a difficult way!".

These little beings, once we called children, are today small monster who have access to a lot of information and, with the help of Wikipedia, they can test whether you have told stories or you have said the truth.

Electronics has been my passion since I was fifteen and I can safely say that my knowledge of the subject is rather broad. If Lorenzo had been ten years older I could properly answer back, but now he was looking at me and was expecting a simple, clear answer.

He thought it was a simple matter, but I had to talk him about the most important invention of the last century, that is the invention of Transistor, the electronic component that is the basis of "all" modern devices such as personal computers, cell phones, smartphones, artificial satellites, without which even the rover Curiosity, wondering on Marth, could not work.

I realized how my duty was to explain those concepts in an easy form to him, even thought I had to postpone to a more mature age the learning of such a important subject of our recent scientific achievements.

A difficult rebus! Credibility and authority were fundamental achievements I had to conquer on his mind to become a professor for him and so to start "teaching" these difficult subjects.

To solve the problem of convincing him about my authority in this matter, I got an idea. In the year 1969 I published a book on semiconductor components entitled "From Transistor to Integrated Circuits" of which I have given a copy to my son, father of my young interlocutor.

Truly the book is a technical text, too difficult for a ten-year-old boy, but if I could find that book among the thousands of paperwork and notebooks stacked in my son's house, I would have solved my first important starting point: my authority.

Together we tried throughout the house and in the end, with great joy of both, we found it.

1969: my first book "From Transistor to Integrated Circuits

I saw his face glowing with a penetrating look, he wanted to know everything and immediately almost ripped off the book to read it and to get the answer he was looking for.

"Careful Lorenzo" I said "this book is suitable for reading and studying for high school students, you have to go to school for at least another seven years to understand it".

But there was nothing I could do. He began to read the introduction, but after a few lines he started browsing for his answer that he could not find.

At that point, however, I had shown him that I knew the subject in depth. I took up my book from his hands and I told him with a certain tone: "The only thing you can understand is the last sentence I have written in this book many years ago and you can read it now!".

The phrase says exactly: "… as always happens in the technique, it is not excluded that this technology, once perfected and made economical, will lead to the creation of devices suitable for applications in other sectors, which are not yet foreseeable".

My intention was to let him know that his iPhone was the result of the development of the Transistor and that in the book I had explained those tings to the students of the technical schools. So we had to start from there to answer his question.

He read it but he did not understand well "what did you mean?" he asked. "Well" I replied with a smile, " I had foreseen your iPhone almost forty years ago, so I can tell you what's inside and how it works, if you listen to me calmly and patiently!".

At this point I decided to spend with him all the free time he had after his school assignments. I knew I was engaging myself in a simplification that for a technical mind like mine is equal to the difficulty of explaining quantum mechanics to a poet.

Fortunately, the cover of my book depicts the photographs of some primordial chips, more precisely the integrated circuits, really widespread at that time, so I could say: "look at those tiny dark chips from which thin wires come out, they are the integrated circuits. They are monsterlings invented over fifty years ago. Since then, they have become smaller and smaller so that millions of them are able to stay inside your iPhone!".

At this point Lorenzo asked me to tell the story of those monsters, how they work, what they eat, if they move, and so on.

His curiosity had become really great!

I opened my book, I showed him some illustrations and with the help of many drawings, I proved how, scaling down more and more those Monsterlings, it is possible to insert many of them in an increasingly smaller space. They eat the electric current coming from the battery, but they are not alive!

He had understood these first simple technical elements, he had had fun, but my lesson had to end at this point, further explanations would have been too complicated.

Back in the tranquility of my home a thought came to my mind: "my nephew is still very young, but how many people know, even if superficially, from where the wonders that everyone today manages easily, come from? It seems to me that no one has been concerned about explaining in a comprehensible and witty way such complex stuff, marvelous achievements of the past century.

This thought forced me to bring back all the many texts and photos that, for my strange passion, I have collected since I was very young, documents that I still owe.

I have a mountain of documents now from which I can deduct what I am writing. These documents contain also a good part of my working life and I am sure some of them are not even publicly known.

I'm sure even the experienced reader will find some unpublished information, and over all, with this book I intend to reveal to all technology enthusiasts what the year 1947 kicked off till today and till a far future.

I therefore invite the reader to relax and read these pages with the promise that at the end of this story he has revived in a short time what the author has lived in many decades. The reader will also get to know the functionality of those electronic devices that have transformed our daily lives.

The Birth of the Transistor

Our story began at the end of 1947 with the emergence of a tiny electronic component. Birth, though silent and uncertain, has indelibly marked our entire life and will certainly accompany the technological development of humanity yet for many years and perhaps for many other generations: I am talking of the Transistor, which we write here with the capital "T" because in this text we want to give him a kind of humanity rank.

The Transistor is the fundamental component of all the electronic equipment that we now buy and use every day with ease.

It's wonderful to think that today inside all our modern electronic appliances thousands, millions and now even billions of these components are working while we hold them in our hand.

They have a heart that is pulsating and they work hard to let us listen music, call a friend and managing other thousands activities at the touch of a finger.

Looking at the following picture that shows the first laboratory specimen we can see how rough was the first Transistor.

At the Bell labs In fact the scientists have put together this embryonic Transistor by using a clip attached as a contact. It's just a homemade device that seems to stand up by miracle ... but it worked!

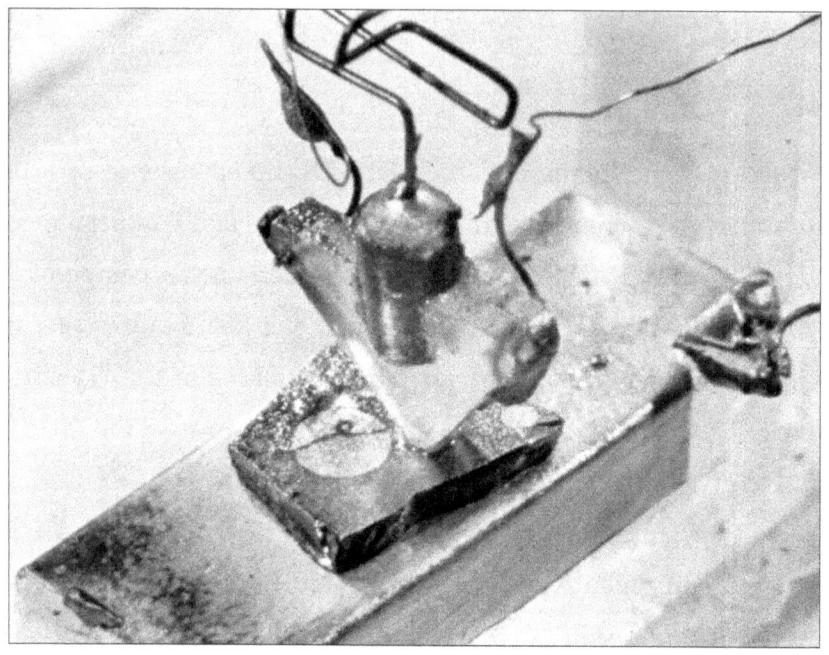

December 1047 - The first Transistor made at Bell Labs

Wanting to give a cute image to this ugly device I thought of creating the funny character on the cover of this book: does not looks like a little monster or "Monsterling"?

And afterwards we will learn why it is also "Semimetallic", in other words, neither metallic nor insulating, but a bit of both.

This small animal, to which we have given the dignity of the uppercase T as if it were human, will accompany us thru this story.

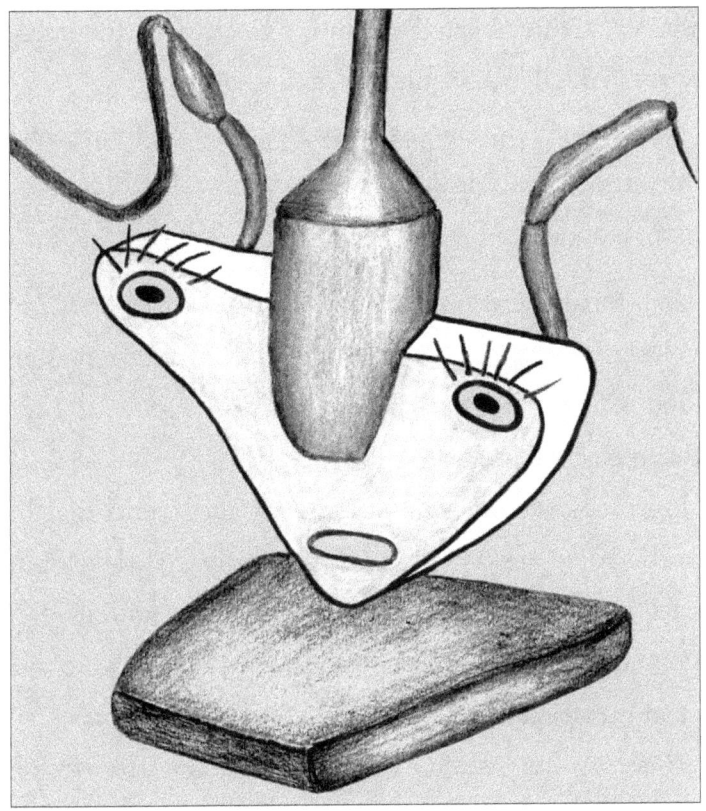

Our Monsterling, funny character of the first Transistor

This little device, to which we have given the dignity of the uppercase T, as if it were human, will accompany us throughout this story.

This Monsterling came to light almost by accident, and at first his parents considered it one of the many discoveries that soon end up forgotten for their apparent fragility.

Instead this creature and all his numerous brothers born afterwards, have grown invading our everyday lives and giving birth to a large number of children, grandchildren and great-grandchildren who are today living inside our PCs, GPS,

Smartphones, artificial satellites and, I might say, they are part of any electronic devilry that surrounds us everywhere.

Here we will follow his story since the beginning and I shall add both autobiographical facts and technical explanation of the most important devices.

I have been fortunate enough to live my schooling period alongside the development of electronics. While student I designed and built many kind of Transistorized devices and once I grew up I continued to deal with electronics.

I also collected a great amount of documents that I studied and I participated as a witness to the industrial developments of electronics components. After completing the studies I had the privilege to collaborate with numerous companies and to visit many factories in Europe and US.

This is the reason why today I am able to offer my testimony of what happened, real facts, some known and some unknown, and all this material now allow me to formulate some reasonable predictions on the future of this technology.

We will see in deep details why all modern electronics has the Transistor as a brick at its fundament and we will learn thoroughly the way and the why. Now let's return to our funny character the "Semimetallic Monsterlings".

The Semimetallic Monsterling

The anthropomorphic character "Semimetallic Monsterling" came to light during one of my conversations with the publisher of a technical/hobby journal with whom I worked in the sixties of last century.

This journal, titled "Costruire Diverte" ("Building is Fun", in English), was directed by Mr. Gianni Brazioli, a sympathetic young enthusiast of electronics like myself who had the initiative to publish in Italy a popular magazine entirely dedicated to electronics.

During one of our numerous meeting in Bologana we were talking about the meaning of the term "semiconductor", the material with which all Transistors are made, and we were just kidding about their tiny size, almost insignificant compared to the vacuum tubes and to their three thin legs that sprang from a cylinder not bigger than an inch.

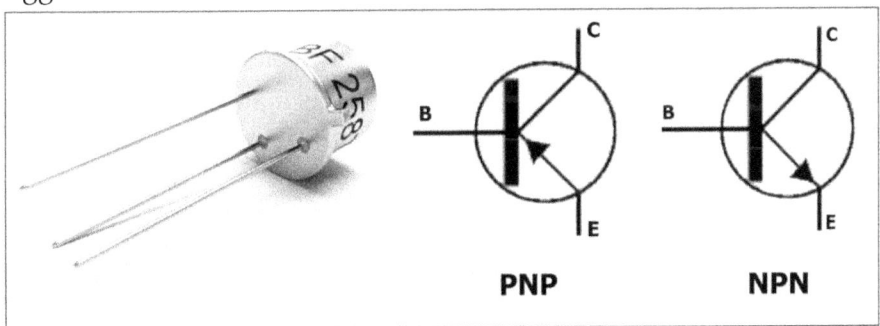

Electric symbols of Transistors

In fact the Transistors are much smaller than the old valves they intend to substitute: the valves, or vacuum tubes, have a beautiful transparent shell through which one can see their interior with its intricate and shiny metallic structure. Transistors are far less attractive aesthetically, in short a cockroach compared to a dragonfly.

From there, jokingly, the "semiconductor" becomes "Semimetallic" and, for its somewhat monstrous appearance, the newly-born Transistor becomes an ugly "Monsterling".

From that day on, we began to use that expression by talking among ourselves. For example when Brazioli was looking for an article I was writing he was saying over the phone: "then, Ettore, when are you going to send me the article about the three Monsterlings that you promised me?", or something like: "Have you seen the new little Monsterling manufactured by Philips?", or "do you know when SGS will start producing high-frequency Semimetallic Monsterlings?", and so on.

We had a lot of fun and we thought to write some articles using the Monsterling as a funny character about whom to build a story, but we thought we could run the risk of being the one to diminish a product we were taking great care and, more, on average, the technicians are not very witty, and they are not even today!

A long time elapsed since then, but when my grandson raised his question, I do not know how, I remembered that fact and I thought of using that funny character as a starting point for my explanations, also because I could immediately see that he liked that name.

A problem will remain when he will have the chance to see this book and I'll tell him that he should wait quite a while to read it and by sure I will disappoint him very much.

I will be forced to explain him carefully that my story here tells of a little monster that does not move does not even eat. I'll see ... anyhow before the release of this book I'll find a way to transform the electric current into a form of food and to invent something at his level.

My Passion for Semiconductors

We were at the beginning of the sixties and after publishing several articles that saw the Transistor making the lion's share, the publisher and myself have agreed that, given the insistent demand of readers, it was time to publish a "Semiconductor Newsletters" by taking advantage of the original technical documentation especially of US origin.

Received the assignment, I immediately started to work, but I found out immediately that it was far for being an easy task.

Internet wasn't yet invented, fax did not exist and to communicate the only ways were by phone or by post. More 90% of the news I needed were at almost 10,000 miles away, in the USA.

Once received the assignment, I immediately started to work, but I found out immediately that it was a quite impossible task.

Internet wasn't yet invented, fax did not exist and to communicate the only ways were by phone or by post. More, 90% of the news I needed were at almost 10,000 miles away, in the USA.

The problem seemed to me enormously complex and not because the companies would not listen to me, companies who, on the contrary, would answer to my enquiries with unexpected solicitude as we will see later, but because with my poor English I could not call in English anyone. More, managing intercontinental phone calls was practically impossible due to their cost and due to many technical difficulties ... at that time there were no ways to dial an oversea call directly.

Even writing letters was for me extremely difficult; I had prepared a general request form but it was extremely limited and the airmail stamp was quite expensive either and so I could just post a few.

I found a wonderful solution only when I discovered the replay cards, that is, the postcards that all the specialized American magazines were inserting inside their magazines and that the reader could send back to the publisher of the magazine who then was forwarding them to the various companies mentioned in the same magazine.

On each of these postcards there was a series of numbers, even one hundred for each postcard and each number did refer to a product or company quoted in the magazine.

Circling each number was like writing a letter to the company to which that number and the request was referring

The publisher was keen to show to his advertisers that his readers were reacting, just as today by clicking an internet ads. For me it was like having found an heaven! With the expense of a single stamp I could send a hundred enquiries!

At that time I think I've been one of the most avid user of that service and, while surely the American readers, mostly technicians and scientists, have been circling just a few numbers in each card, I was circling practically any number.

So, at some point at home, where I was living with my parents, a mountain of envelopes and documents began to arrive daily from oversea and from the most diverse companies. I must to say that these companies were extremely generous by delivering to me manuals , application notes, schematics and so on, matter very useful to their industrial customers ... and to myself too.

By sure my generous senders did not imagined that all that work was destined to a young Italian student, however, I remember the astonished face of the concierge guard whenever she was handing over me those almost five or six pounds of envelopes almost every day.

My home was flooded by a lot of documents written in English, documents that I barely could decipher and that no other person could help me to translate because I was the only one in the family with a rudimentary knowledge of English and electronics.

I am sure that at that time, in the middle of the cold war, even the CIA was concerned about my traffic, as I did not distinguish between civilian or military companies and I was enquiring all without distinction!

Among the document that I received I remember a strange and interesting letter from Western Electric, very courteous, that was informing me that they could not satisfy my requests because their entire production was absorbed by governmental commissions and therefore I had to accept their denial. Too bad that I did not keep that letter ... however I accepted willingly their apologies.

Thanks to all this work, I was able to fill out an extremely updated monthly newsletter, full of technical information about the many new products that American companies announced daily.

I did not overlook either Europe and I have to say that with two companies I kept a close correspondence, namely the Dutch Philips and the Italian SGS.

Besides documenting me massively, SGS (Società Generale Semiconduttori, Agrate, Milan) made me one of the most pleasing gift: a box of containing twenty germanium Transistors and a silicon Mesa power Transistors produced in collaboration with Fairchild Semiconductor, at that time partner of SGS.

This Mesa Transistor could work at very high frequency and represented an absolute novelty on the market. Thank to its technical specs I became the first to publish a project of a powerful transceiver completely Transistorized. By the way, I still have with me that transmitter.

Then my interest was in fact not only to report news in my newsletter but also to self build electronic devices.

At home I was filling my table with all the samples I could find and with which I was realizing various projects that a couple of publishers were happy to buy so I earning good money for my age.

I also realized a small transmitter to copy the tasks at school, requisitioned by the professor the first time I tried to use it on the third year of high school. I believe that the transmitter is still part of the small museum within my Zaccaria school in Milan

Before to proceed I must open a parenthesis here to explain how my interest in electronics started. It has been a very specific fact that turned on my light bulb and I think it's interesting for the reader that I describe what happened to me because it explains how important is to discover tendencies in childhood, whether they are technical trends or artistic trends.

For this reason, I will describe this adventure from the very beginning when, at the age of fourteen, it came to my mind a little book that has awakened in me a profound curiosity.

From that moment on, I started a journey finding so many difficulties that I have passed with tenacity. I will explain how I felt the need to try to build what I was reading and how strange terms like "Galena" and "Hertzian Waves" created a curiosity in my mind that I had to satisfy and he joy for my achievements.

Galena and Hertzian Waves

My adventure began in the year 1956 when at age 14 I bought a small booklet in a Sunday sale at my school. This book was describing some electrical equipment that a young reader could self build. Incredibly I still have this booklet now, over my table at my left, as I write these lines.

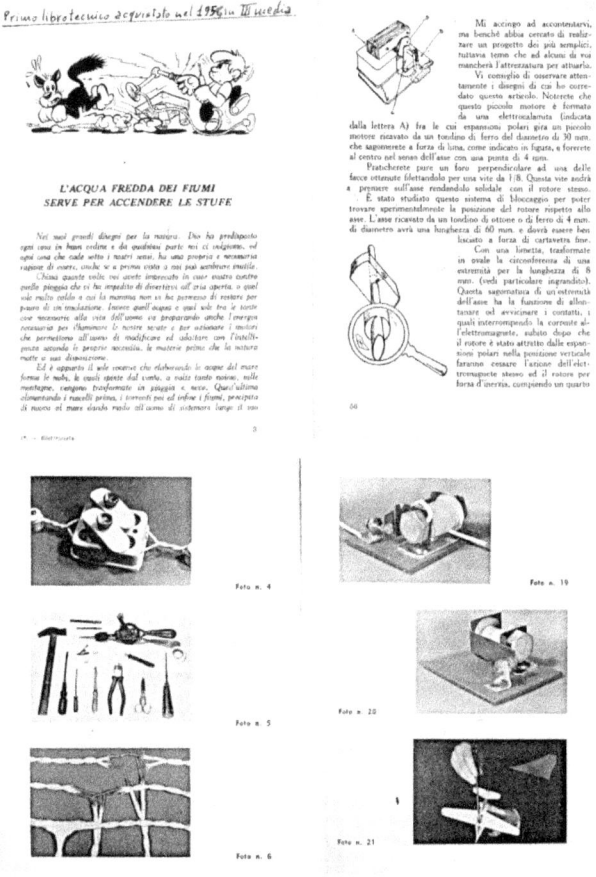

Some pages of my first technology booklet I bought at 14 years old

The author is a father Barnabit, a priest professor of the Institute I was attending and that wrote this book using a language that was quite understandable for a boy of my age and with naïf pictures showing how to build an electrical motor, a battery, a radio, etc.

The booklet explains how electricity is produced, how to build working tools and it concludes with a final philosophical consideration that I translate in English here below:

"CONCLUSION

To draw a conclusion after all this series of topics discussed, after this amount of work, I confess, I find it difficult.

I would like to take you to a small consideration: all that we have done, all that we have studied is nothing but a fleeting glance, that an elementary experimentation of phenomena of a cyclopean force of nature.

One of those mysterious energies that man tries to dominate without success, one of those wonderful gifts offered by the Creator to the human intelligence, to engage him in research, to strengthen himself in conquest, to humble himself in the face of the immensity of his works .

We are still bowing our heads in the face of such a demonstration of strength, once more we join our voice with that of the Holy: "Blessed be my Lord with all your creatures ....".

Indeed, despite the fact that the language of the book was pretty basic or at least it seems to me by reading it now, I remember that it took several days to myself, if not months, to understand certain statements and certain schematics.

The booklet consists of several chapters which, starting from the explanation of what electricity is, comes to the last chapter, the fifth, which contains the explanation of mysterious "Hertzian waves" and that is exactly what the chapter intuitively means.

The strange "Hertzian" adjective, which seemed to me more a Martian word than a scientific one, immediately attracted my attention, and in fact, ignoring all the previous chapters, I went straight to reading what these astonishing waves were like.

When I started reading the chapter, I was immediately surprised by the fact that the chapter began stating: *"dear young reader, I'm sure you jumped at this point without reading the previous chapters ... attracted by the curiosity of the title"* ... this author's anticipation left me puzzled and I wondered how he could have predicted my move so well, as if he had read my thoughts.

Continuing reading I am advised to go back and read the previous chapters if I want to understand that chapter as well, notice that I do not intend to follow, acting exactly as my grandson who grabbed my book and wanted to read the answer to his complicated question at once.

I read the chapter all of a breath, understanding quite nothing, but I got attracted immediately by the practical part, although about the blessed Hertzian waves I only understand that they are electromagnetic waves discovered by a certain Mr. Hertz and from whom they have taken their name and that they move in the space who knows how.

What do I find in this chapter that ignites that spark of curiosity that will have remarkable and positive consequences in my years to come? Something that would have prompted me to deepen the subject and make me unknowingly to participate in a research that would, for a long time, be the basis for my future studies and my work, once graduated.

The booklet report a picture that explains how to build a Hertzian wave receiver and which scheme I scanned here below now.

I understand that for a technician this scheme is very strange, but I must say that the author had seen just how to explain it to young readers like myself.

More than an technical scheme this is the drawn of the pieces to be assembled: note the coil with the various sockets, the variable capacitor, the ground socket connected to the water tap and, exaggeration, the antenna connected to the top of a bell tower.

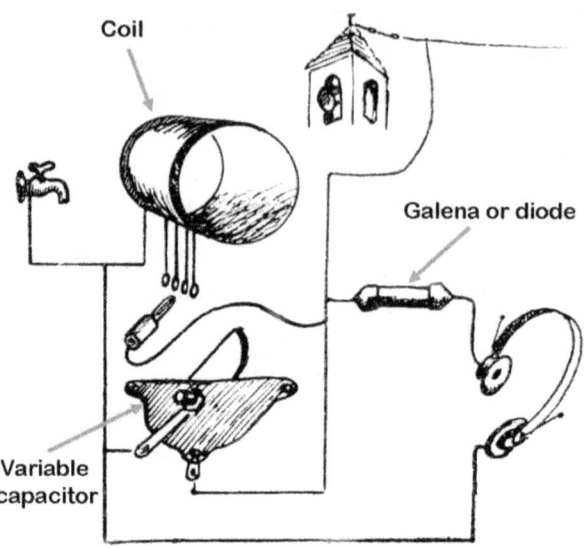

Primordial construction scheme of Crystal radio

The correct equivalent electrical scheme that a student of technical subjects should be able to interpret is as follows:

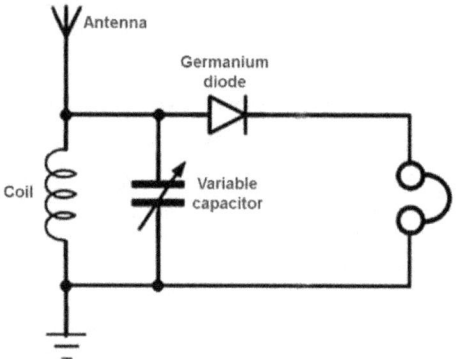

Technical schematic of a crystal radio

And sincerely, I learned to read and interpret these kind of diagrams only after many months, when they were explained to me by some adult person and who, with patience and for me in advance of a few years of school time, allowed me to access and understand those basic elements of electronics with which I could accomplish what we will see.

Hertzian Waves Receiver

As I said earlier, at that age I didn't matter to understand the physics behind the technology and even less how everything works, while the idea of being able to build an instrument like this one has been exciting me in the highest degree: to build that thing through which to hear these Hertzian waves was a must for me!

It was for me a bit as if a kid of today time could self-build a Star Wars light-saber able to function as seen in the movies. What the construction scheme suggested to the reader was simply how to build a small crystal radio and with which he could hear some of the medium wave stations through a headphone.

The booklet explained quite well to me how to link together the various components and how to self build some of them and the game seemed to be easily done. That tiny radio could work with no battery or any kind of electric power and could receive even very distant radio stations, so the booklet said.

Imagine how excited I could be with the idea of being able to accomplish a task that would allow me to hold in my hand something that would work like the giant valve radio that was standing majestic in our living room. Only Dad could turn on the big radio every evening and, with devotion, to listen boring radio news ... and then been turned it off! A TV set was not yet arrived at my home and only then began to spread into the Italian homes.

My basic task consisted in following step by step the instructions drawn in the booklet and to connect the various parts that the picture in the booklet was showing, something feasible for someone like me who did not have any technical knowledge.

At this point, however, I found myself between two new difficult terms: "Hertzian" and "galena", which together encompassed the whole mystery of that device that now "I wanted" absolutely to build.

Hertzian waves were no problem; I had not to build them, but the galena, a substance totally unknown for me, was a real mystery. Only after many months I found out that this substance is a fairly common natural mineral in the form of lead sulfide and that, under certain conditions, can operate as a rectifier of alternate electric current.

I think that today only a few know how to build a radio receiver with no electric power supply and no electronic parts ... in an emergency situation, still now, I could build a radio crystal, even with no galena and no diode ... but I don't tell you!

I must mention how the crystal radio has been the most popular radio receiver used for decades at the turn between the nineteenth and the twentieth century.

We are in the year 1956, I am fourteen and I begin the arduous task of assembling a self-built coil, a variable capacitor and a detector in the form of a galena crystal as the picture in the book is showing and so to achieve an audible signal on my headphone.

While today may seem a very easy task, to me at that time was a quite hard job. I knew nothing at all on top of the little chapter in the booklet, but the anxiety to build this blessed radio made me to overcome any obstacle.

To build the coil was easy, by winding wires of copper around the cardboard tube removed from the inside of toilet paper. I found fairly quickly the variable capacitor and the headphone at an electrical shop near home, but .... the galena?!

Damned galena! No one could tell me where to find it! The task was huge for me, without galena my Hertzian waves would remain out there and I could not hope neither to see them nor to hear them!

No electrician or shopkeeper that I visited did know where I could find a galena. Of course, Internet did not exist at that time and my parents, even if they tolerated these researches of mine, were seeing me wasting a good amount of time borrowed from my daily studies.

The school examinations were close and, not only they did not cooperate, but constantly they were checking my homework ... how boring!

But now I could not stop: galena was my obsession, I had to find it! And so after a month of frantic searching, someone suggested me to try to visit a shop of electrical trinkets, quite far away from home.

You can imagine how fast I rushed to that place and right there I found my beloved galena.

The place is a small shop, a bit run-down, but full of small wonderful objects, plugs, wires, capacitors, resistors, valves, used equipments.... in short, a paradise!

The shopkeeper at the beginning does not take me seriously: the galena is no longer used since many years and what the hell can make with it a little boy with milk on his lips?

Fortunately, after some talk and a few questions, I explained my goal and I conquered the confidence of what seemed to me an elderly shopkeeper (probably he was thirty-five years old!).

He then goes into the back room and from there he emerges a little later with a cotton swab on which rests a tiny silver crystal, gleaming and he says "this is the galena you are looking for, you must insert it into your circuit and you must use a metal pin to connect it to your circuit and you must move gently the pin till when you will hear a sound on your headphone".

It puts everything in a transparent plastic box, just like jewelers with their precious stones and hands it over to me with the delicacy with which it would handle a valuable diamond.

And in fact for me sparkles more than a diamond and is more precious than the most precious gem of the earth.

I still remember his other recommendations ... do not touch it with your hands .... you have to put it in a small container where the mobile metal pin is in contact to its surface ... move the pin as gently as possible and search for the zone where makes you to listen to a radio station .... you need to be very, very patient.

I received that treasure with the utmost happiness and I paid with those few liras in my pocket and, happier than ever, I returned home, anxious to complete my construction.

Galena mineral and crystal radio detector

I will never forget my great joy, mixed with wonder, when I have inserted the galena in my primordial radio, built inside of my dad's Turmac cigarettes cardboard box, and by moving the metal contact suddenly I could feel and clearly a radio station, exactly Rai 1: what a victory!

I felt like a magic that a small box in one hand would allow me to receive two radio stations, really just one well enough. What could I wish more? I had my antediluvian iPhone ... and it was working with no batteries!

I could not understand how it was possible that a piece of sparkling mineral with its precarious contact could receive far away Hertzian waves and so clearly and intelligible.

It looked to me as a great mystery how the forces which were contained in that piece of mineral could produce all that!

But my story does not end here; soon I'll tell my readers how this story about the galena ends up.

In fact the reader might believe that the small radio in the cigarette box was for me a sufficient accomplishment and that I would have stop my research pleased of the result. But it has not been so; just as today we are used to move from iPhone4 to iPhone5 and so on, never satisfied, I have experienced this syndrome and soon I tried to improve my device.

Germanium Diode: unexpected Solution

As they say, hunger comes with eating, and though I was very happy of my result, soon I'll find out all the limitations of that radio box and so I started a new search for overcoming them.

In fact I soon began my run to improve my underperforming radio. More, my curiosity drove me to look for understanding everything about that primordial and natural semiconductor in which, I thought, Hertzian waves were concentrated and the mysterious force that allowed me to hear far way radio stations.

A curiosity that pushed me to follow the development of the semiconductor components industry till today.

I can now offer to the young reader a vision of the technological phenomenon that I, unknowingly, had the luckiness to follow day by day from the beginning, starting from the Transistor to the today Artificial Intelligence.

It a pleasure for me to accompany the reader to the understanding of the phenomena that then took me years to understand, I mean those phenomena that are the basis for the development of modern electronics of which the Transistor is the grandfather.

Let's go back to my galena and to my radio. For a long time I could not find any explanation on the origin and functioning of the galena, indeed I just found that, although it was known its use, it was still a mystery how galena would operate: i.e., it was known its application but not the theory behind it.

For the convenience of student readers I summarize here below what Wikipedia says today about this mineral.

"Galena is the main ore of lead, used since ancient times. Because of its somewhat low melting point, it was easy to liberate by smelting.

Galena deposits are found worldwide in various environments. In the United States, it occurs most notably in the Mississippi Valley type deposits of the Lead Belt in southeastern Missouri and in the Driftless Area of Illinois, Iowa and Wisconsin.

Galena is the official state mineral of the U.S. states of Missouri and Wisconsin; the former mining town of Galena, Kansas takes its name from deposits of this mineral.

Galena belongs to the octahedral sulfide group of minerals that have metal ions in octahedral positions, such as the iron sulfide pyrrhotite and the nickel arsenide niccolite.

The galena group is named after its most common member, with other isometric members that include manganese bearing alabandite and niningerite.

Galena is a semiconductor with a small band gap of about 0.4 eV, which found use in early wireless communication systems. It was used as the crystal in crystal radio receivers, in which it was used as a point-contact diode capable of rectifying alternating current to detect the radio signals.

Making such wireless receivers was a popular home hobby in Britain and other European countries during the 1930s. In modern wireless communication systems, galena detectors have been replaced by more reliable semiconductor devices"

Since the early years of the last century, when the galena radio was the only and popular way to listen a transmission, no one knew the physics of operation: only many years later, when the solid-state physics would reach its maturation everything would be clarified.

I too, many years after my first experiments with the galena, studying the operation of the Transistor and the associated phenomena, I could finally solve this mystery, a mystery that later will be revealed also to the reader".

Going back to my crystal radio, I was walking around my house with a bulky headphone on the head connected to the cigarette box from which a long wire as antenna was lying behind me on the ground.

I had to be very careful not to shake the box otherwise the metal point which rested on the galena would move and I could say goodbye to my listening. Even a small shake of the box, and often was happening, I had to gently move the tip until I could get back my transmission.

It was an operation that sometimes could last several minutes, but the biggest problem for me was late at night, when parents were forcing me to go to bed. Under the covers, in the dark, instead of sleeping quickly as they would have liked, I have been trying to listen to the first radio channel by moving the contact of the galena, but in the dark was a feat often impossible.

To find the damn point of good listening I should turn on the light of my bedroom but so I would have revealed to my parents what I was doing and after many attempts in the dark, I was forced to give it up.

It has been to overcome this problem that I was driven to make a leap that will lead me to the discovery of new electronic components and I invite the reader now to do this jump with me by passing to the use of a germanium diode in place of the galena.

Germanium diodes are part of the "solid-state" modern electronic family (diodes, Transistors, chips, microprocessors, etc.) made by semiconductor, solid materials, to distinguish them from the vacuum tubes; the most popular semiconductors are germanium and silicon.

Semiconductors have the particular characteristic of a moderate capacity of conducting electric current, from which their name, and this conductivity can be adjusted to the desired value needed in the production of solid state devices.

In the fifties and sixties it was customary to call all electronic components based on semiconductors, like diodes and Transistors just with the name "semiconductors". From now on with that term we will mean all the electronic components being part of the solid state family and that have been part of my research.

As I said, wanting to improve my radio so I started a new research on the possibility of replacing the uncomfortable galena to eliminate the instability that forced me to continuous manual adjustment of the contact point.

To start let's go back to my nightmare! I returned to the only source of information I had available, the shopkeeper friend that sold me the galena few months before and that provided me with all those explanations and that put me on the right track.

I exposed my problem and asked him if there was a more stable galena, something I did not need to move that damn point contact.

It 's then that I discovered that galena was no longer used since many years and that he was keeping some pieces for old people who love technology archeology and he added that he have sold to me the galena only because I have been asking to him exactly that.

He added, a bit surprised, that by sure he could not imagine that I was doing everything by myself, but he thought that behind my request there had to be or my father or, more likely, my grandfather, otherwise he would have suggested not to use the galena, but another more modern component called "germanium diode".

And so, always thanks to this shopkeeper, I discovered the existence of some very recent components which, when inserted in the circuit in place of the galena, it would be doing the same function but in an absolutely stable way: nevermore unstable point-contact.

These tiny devices, called germanium diodes, consist in a glass cylinder of half a inch from which two electric wires protrude.

He told me that they came to the market very recently and that they could replace some vacuum tube in the radio market. He added that the valves have been invented at the beginning of the twentieth century, after the galena.

Now I had another big problem: the diode was far too expensive, well more than the few liras in my pocket, roughly two weeks of my student salary, so I had to wait to accumulate that large amount of money before to return. I came back after a couple of weeks and I purchased that precious germanium diode of which I still remember the name: Philips OA70.

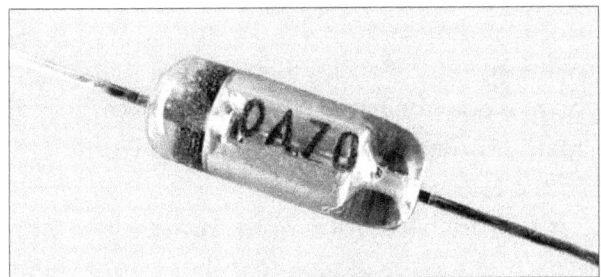

Germanium diode

Back home I immediately inserted the diode in the cigarette box and I could suddenly listen my radio stations over the headphone and I could shake the box without losing the connection ... what a marvel!

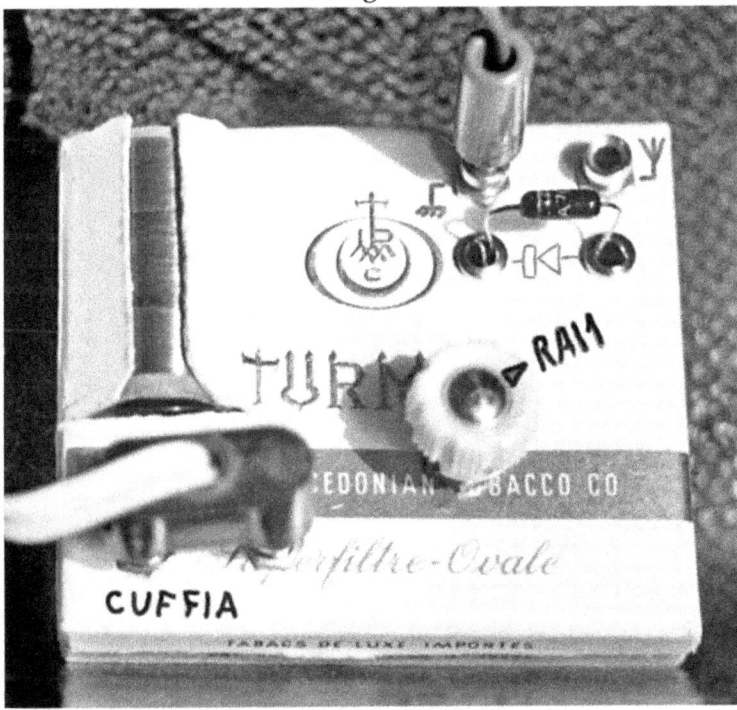

My crystal radio with germanium diode

I should say that I was listening that radio channel just for the pleasure of the novelty, but the only station I could listen was Rai 1, a real nuisance and often incomprehensible to me: news and politics far away from my reach and even the music was mostly classical and had only the result to conciliate my sleep.

Anyhow I got my hands on the first semiconductor of my life: the germanium diode and the Transistor would follow soon as we will see in the following chapters.

In writing this book I imposed to myself to provide young readers how each device or component that I quote here works and in a way the most could understand.

.

Let's find out how the Radio works

Following the spirit of this text that aims to make simple and interesting concepts sometimes not simple, we'll use a horse and its rider and if the reader will follow my reasoning at the end everything will be clear.

You should know that the signal arriving at the antenna, the damn hertzian waves, cannot be heard as they come. They are high frequency electromagnetic waves, not audible by our ears.

What the the diode does? Simply it pull off from the hertzian wave what the technicians call the "audio signal". This audio signal reaches the headphone that, by vibrating, moves the air and create a sound that our ears can listen.

We can represent hertzian waves like a horse carrying a knight, i.e. the audio signal, and that together they depart from the radio station that is broadcasting. When the horse with its knight reaches the first hedge (see figure), i.e. arriving at the antenna, they find the first obstacle represented by the variable capacitor and the coil.

Overcome coil and capacitor (we'll see later how and what they do) they meet the diode who order the knight to get off his horse if he wants to pass; i.e. the audio signal only his allowed to move towards the headphone.

In the vein of explanations with horses and knights, let's use the example also to understand what that coil and that variable capacitor are doing there.

Yes, because at the end of the day I also wanted to understand what those two components were for. And here's the explanation: the Hertzian waves that arrive at the antenna are actually more than one, i.e. the antenna collects the carriers emitted from all the radio stations that are located in the surroundings, for example in my case Radio 1 and Radio 2.

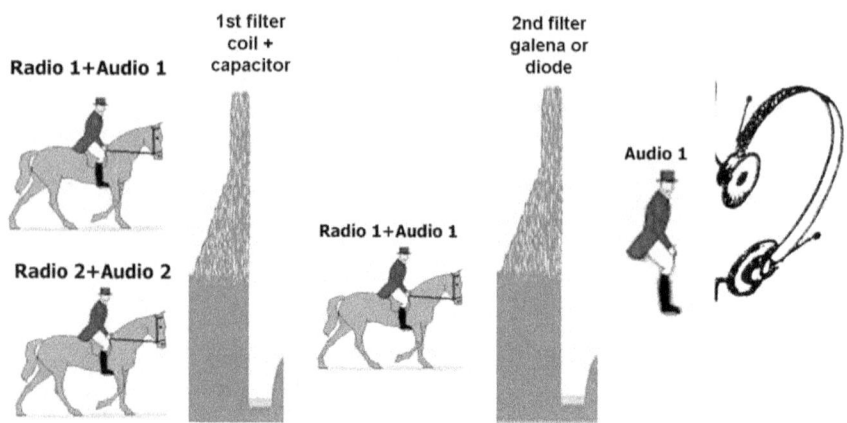

Simulating the reception of a radio station

Now, if you want to listen just Radio 1 you must filter all the other incoming signals before the diode. And this is precisely the function of the capacitor and the coil which filter unwanted stations and that, by turning the knob of the variable capacitor, allows only to the selected station to pass. This operation is called "tuning" of the signal.

With my antediluvian crystal radio I was receiving two stations, Radio 1 and Radio 2, and by turning the knob of the cigarette box to the second notch I was tuning my radio to the faint Radio 2 and blocking Radio 1.

I had thus completed, after much time and study, the knowledge of the functioning of my entire radio equipment, namely and summing up, that the antenna was receiving all the hertzian waves, said carriers, that the variable capacitor and its coil allowed to pass only one horse and his knight, my Radio 1, and that, finally, the diode was stopping just the horse and allowing the audio signal of Radio 1, that I selected, to reach the headphone ... and so I could listen my boring radio station.

Frankly speaking we just learned a subject that I studied at the third year of my course at the Polytechnic of Milan, a quite difficult subject, to be sincere, of the communication course.

By sure I did not answer in terms of horses and knights to my examining professor, anyhow my grandson has appreciated this explanation of mine and I am sure he will remember it much better of what I remember of the serious theory I learned almost fifty years ago.

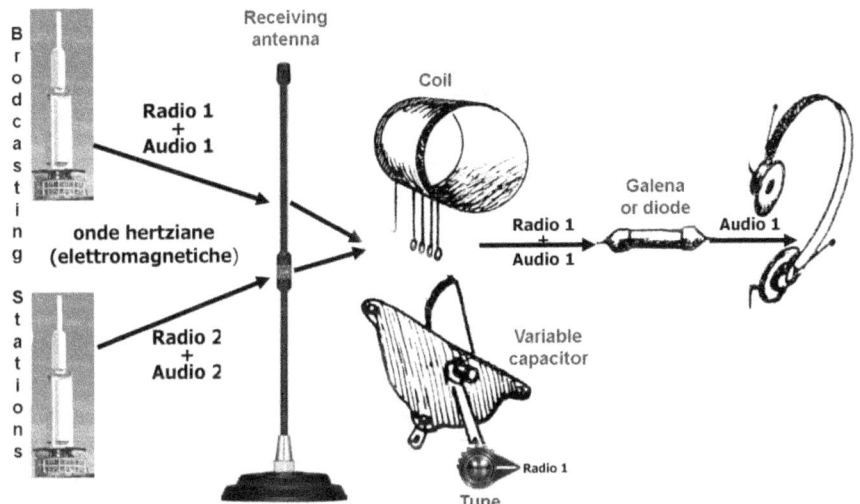

The previous simulation with real components

Having said that, I spent a few months with that radio that was working with no battery and a long wire as antenna. A radio that was allowing me to listen just a boring radio station and soon I was feeling a deep dissatisfaction.

I started to hate that single station and its classic music, it was really far away from the taste of a teen ager like I was!

With the hope to find more radio stations I was comparing, with a sense of humiliation, my little box with the giant five valve radio standing in the living room. I remember the embossed brand "Telefunken" on its large brown wooden furniture.

How an old Valve radio appeared in 1956

I had to extend upward my arm to reach its knobs because, given its weight, my parents had placed it on the top of a small wardrobe with two doors, and so I could turn the knobs of volume and tune.

After turning on that big radio I had to wait few minutes before the loudspeaker cold emit any sound. The reason of that delay was due to the five valves that needed to warm up before they could start working. When, finally, the powerful sound was coming out from the big loudspeaker, the radio lit up like a Christmas tree . My poor radio was in evident inferiority!

On the front panel there were four knobs, volume, tone, tune and one knob that was selecting short, medium and long waves.

By turning the tune knob I could find many Italian radio stations and turning the fourth knob to "short waves" I could even listen distant stations broadcasting in strange languages, languages to me totally unknown.

It was evident to me that those five lighting valves inside the wooden box had to have some role in this super power. On the other hand it was clear that the signal coming from space to my antenna could not be much different from the one that came to the antenna of the giant radio that was right near my self-build radio.

I put the two antenna wires, mine one and the one of the valve radio, in parallel just a few centimeters away, but the situation did not change, in that confrontation my radio seemed a failure.

It took me months of reading to understand a simple fact that probably for many readers is obvious. In my crock without battery there was no signal amplifier, while the radio with five valves was receiving the same weak signal from its antenna, but it was multiplied by the valves, or rather amplified, thousands or millions of times.

I was receiving on my headphones the very weak signal as captured by the antenna and rectified by the diode. And only those captured electromagnetic waves that were powerful enough could have the capacity to move the air inside my headphone; in my case, just one very close radio station could do that.

So I discovered the concept of "amplification" of the electrical signals, i.e. devices powered by electric current able to receive on the one hand a weak signal and to reproduce it at the output, equal to itself, but much more powerful.

And that was the function of each of the valves inside the giant radio in the living room. The first valve receives the signal from the antenna and amplifies it, say fifty times, then it pass its output signal to the input of the second valve that amplifies a second time fifty times and so on.

Now it was clear to me that my radio could not compete with a radio capable of amplifying the antenna signal even millions of times.

Understood what I needed, my new task became really complicated: how could I put inside my little box of cigarettes even just one of those big valves to boost my signal?

And, more, I should connect my radio to the power outlet and the wire would have been clearly visible and then, I reasoned, I am sure my radio in the evening, soon or later, would have been confiscated by my parents to prevent me from listening it hidden under the blankets.

No, I had to find some deviltry which could resolve the issue, something that would behave like valves, but that would work on a portable battery.

And so I started a new research to solve my problem and it is with this new research of mine that I discovered that lovely Semimetallic Monsterling, the Transistor, which could take the place of vacuum tubes. The Transistor is a very small device and does not need much electric power to run, a tiny battery is enough.

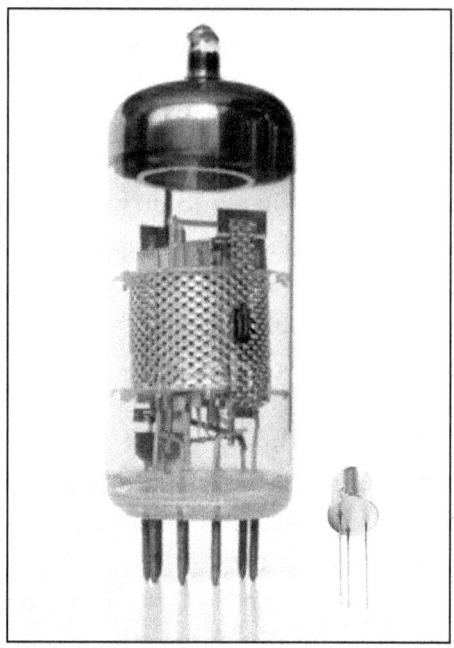

A thermionic valve (vacuum tube) compared to a Transistor

Looking inside the transparent glass bulb of a valve one can see its complicated structure, with all those strange metallic parts whose purpose is to govern the electron flux emitted by a tungsten filament heated up by the electric current.

The amplifying effect of the valve may be explained by saying that the electrons get accelerated by the external power supply and forwarded to the output.

Electrons, in a Transistor, move in a solid body and there they gain force from an external power supply.

Transistor Invention

Let's abandon for a moment my personal story and how I've pretty much solved my problem of increasing the power of my radio, and let's talk about the history of the Transistor and how was born and how it works.

To this end I will summarize what I have studied and understood many years later and I will use a simple language understandable also by non-technical. I consider this topic a fundamental premise to understand the following developments.

We have to move to Stow Lake, in the USA, at the Bell Laboratories, namely shortly before Christmas 1947, where two scientists, experts in solid state physics, were fumbling around a small germanium crystal, such as that one in my germanium diode with which we dealt in a previous chapter.

This germanium crystal has been suitably prepared and connected to three wires. It was a small sample handmade and very primal, one of these wires was even made with a paper clip bent, as can be seen in the photo at the beginning of this book.

These two scientists, John Bardeen and Walter Brittain, were supposed to report to their boss William Shockley that assigned them that experiment. All three will be honored by the Nobel Prize in 1956, precisely for the invention of the Transistor.

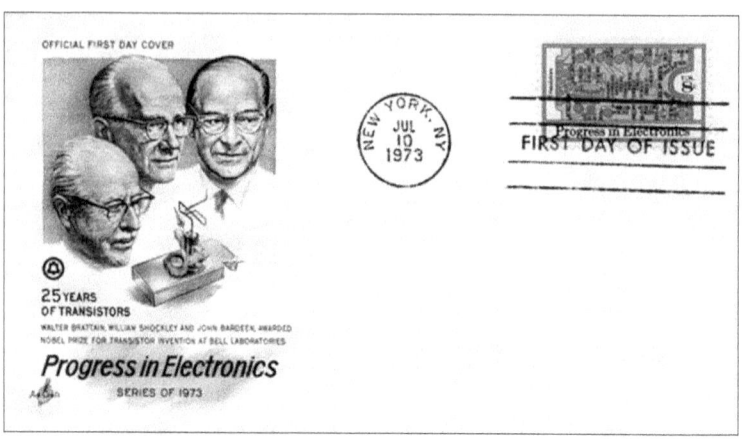

Commemoration of the 25th anniversary for the Nobel Prize of the Transistor

Wonder of wonders and great gift (it was almost Christmas), these scientists discovered that modulating the electrical current of a contact, the same modulation amplified was appearing at another connection!

Precisely the same type of amplification that I had discovered to operate in the valves of the gigantic home radio. But now that phenomenon occurred in a tiny semiconductor crystal of germanium rather than in a vacuum tube illuminated by a tungsten filament similar to the filament in a light bulb of ancient memory.

The first solid state amplifier was born and will give rise to a new generation of very numerous solid state components that soon will substitute in many applications the valves invented at the beginning of the XIX century.

Bardeen, Brittain and Shockley will call this new amplifier "Transistor", and later on it became "point contact Transistor" to distinguish it from the "junction Transistor" invented in the 1948.

This new device, even ugly and fragile, for his parents was an unexpected hope, hope that will far exceed every possible expectation of its inventors.

My discovery of Amplification

Before continuing the historical analysis of the discovery of the Transistor and what followed it, let's go back to my crystal radio and, as you might imagine now, my solution could only be the Transistor, of which I had to find urgently a few to complete my construction.

Looking around I discovered that some germanium Transistors were on the market and that could be bought at some specialty store that I started to look for as at the time of the galena, but now I could move more nimbly.

I was now able not only to understand circuit diagrams but also to modify them to fit my needs so, once bought three Transistors (for the experts they were germanium Transistors of "alloy junction" type manufactured by Philips and branded OC71) I began to study how to build an amplifier to be connected at the output of my crystal radio.

In my mind the three Transistor amplifier was just a first step: I thought that if five valves allowed to receive radio stations thousands miles away I would have to design and build an amplifier with the maximum number of Transistors possible, at least five or six, or even ten.

But I soon discovered that there were two major limitations to maximize the amplification. The first was the cost: if the germanium diode seemed to me quite expensive, the cost of each Transistor seemed to me stratospheric, each costing ten times the diode!

The second limitation was technical and I would have discovered it much later. In fact you cannot cascade an unlimited number of Transistors, or even valves, and amplify the signal an unlimited amount. The reason being that while you amplify the radio signal coming from the antenna you amplify also the electromagnetic noise which is accompanying the signal.

So if you amplify too much, the noise ends up choking the good signal. More, not only the noise arrives from the antenna, but also each electronic component generates its own noise, called thermal noise, which is why we must stop at a certain number of amplifier stages.

This is why the radios on the market were at a maximum of five valves (many had just three) and the famous Japanese Transistor radios contained five or six Transistor at most.

Thanks to a financial help of my parents, obtained thanks to good grades at school especially in science subjects (with Latin I was not getting along too well), I managed to buy those three precious Philips Transistors.

With these Transistors I put together a small three-stage amplifier, powered by a 1.5 volt tiny battery and able to amplify the output signal of my radio.

Despite many attempts I could not put this amplifier into the cigarette box, even though it was very small, it was bumping against the variable capacitor preventing it to turn.

I then chose a solution that will prove very funny: I had received as a gift a nice fountain pen with his little cardboard box shaped around the pen, long and narrow like the shape of the three Transistors connected to each other and the small battery
I drilled a small hole in the base of the box and I inserted a small

jack connector for the headphone. On a side of the box an electrical wire was connecting the cigarette box to this amplifier.

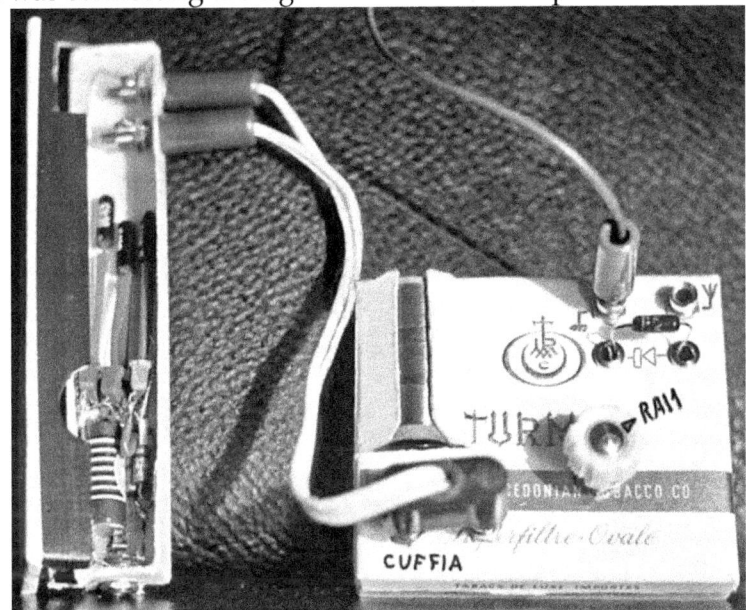

1957: my self-built three-Transistor radio

Everything worked great! I could listen many Italian radio channels ... well, many ... the few that were broadcasting at that time and I marked each one around the knob on my cigarette box.

Now I owned something really working well: my first medium wave Transistor radio.

As far as short waves were concerned nothing I could do: the Transistor commercially available were capable to amplify audio frequencies but for the high frequencies needed to boost short-waves I'll had to wait few years to see them on the market.

Germanium Transistors available could not yet compete with the valves both in terms of power and of frequency.

It will take years for Transistors to be able to replace valves in almost any application and I should say that even today, in the twenty-first century, there are applications for which only the valves are capable of delivering certain performances.

I like so much the set of cigarette box connected to the pouch fountain pen box that I decided to photograph it and to deliver my project to a technical magazine, for the truth, without the slightest hope that it could be published.

You can imagine my surprise and joy when a few months later, buying that magazine on a newsstands, I saw my article cited on the cover and the whole project and my photos published.

And my surprise did not ended, after a few days I got from the publisher an envelope containing a check for ten thousands liras.

I had never seen a check, but most importantly, I had never handled such a large sum and it was also the largest party in the family and at school when I showed the magazine.

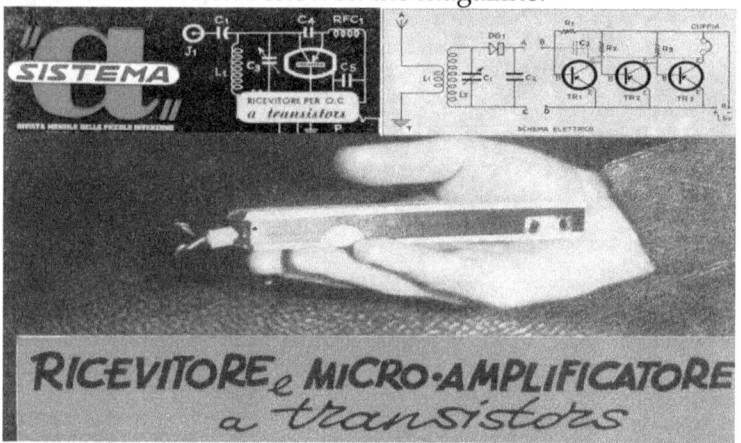

Feb 1959 - My three Transistor radio project appeared on the cover of a technical magazine in Italy and became a great success both at home and among my schoolmates and professors. The payment from the publisher was a great surprise

At school I got the nickname of "Archimede Pitagorico", a name derived from the Walt Disney Italian fiction character that in English is called "Gyro Gearloose".

From that success, definitely unexpected, I felt pushed even further to investigate the world of Transistors and semiconductors and to continue publishing electronic projects and, more, to get well paid too.

How the Name Transistor came out

I often read in the technical literature that the name "Transistor" was decided by its inventors as a contraction of the words "transfer" and "resistor", to mean that it is something that lies between the two concepts, neither conductor nor resistor.

I discovered that the choice came from a different path as we will see.

A document issued internally at the Bell Laboratories entitled "Terminology for a Semiconductor Triode - Committee Raccomandation" and dated May 1948, of which I have a copy that I reproduce here, was aimed to a group of Bell technicians and it was inviting them to vote about the choice for the name of this new device.

The document was asking to choose from five names: Semiconductor Triode, Surface States Triode, Crystal Triode, Iotatron and Transistor.

As we know now, the majority chose "Transistor": so a democratic choice and not that of the inventors only!

By reading the first patent filed on February 1948 by Bardain and Brittain I could see that there is no mention anywhere of the name "Transistor" and this device is cited there as "Three Electrode Circuit Element".

BELL TELEPHONE LABORATORIES
INCORPORATED   D.   CONFIDENTIAL

COVER SHEET FOR TECHNICAL MEMORANDA

SUBJECT: Terminology for Semiconductor Triodes - Committee
Recommendations - Case 38139-8

COPIES TO:
1 - Dept 1000 File
2 - R. Boyn - Case File
3 - R.K.Potter
4 - J.R.Wilson
5 - G.W.Gilman.
6 - J.W.McRae
7 - H.S.Black
8 - H.C.Hart

MM-48-130-10
DATE May 28, 1948
AUTHOR L.A.Meacham
C.O.Mallinckoodt;
H.L.Barney

Surface States -
Terminology

BALLOT

Designate by the numbers 1, 2 and 3, the order of
your preference for the names listed below:

_____ Semiconductor Triode

_____ Surface States Triode

_____ Crystal Triode

_____ Solid Triode

_____ Iotatron

_____ Transistor

_____ _____(Other suggestion)

Comments: _____

Signed _____

Please return this ballot to Miss G. R. Callender
in 1A-323 at Murray Hill.

Original document dated May 1948 Murray Hill Bell asking

This shows how it is difficult to attribute a new and appropriate name for something that did not exist before! However I must say that the choice was perfect.

Transistor: Invention or Discovery?

To better understand what happened at that Bell lab and how those three scientists have been able to obtain the equivalent of a thermionic valve in a tiny germanium crystal we must look back in the history.

In fact it has been since a long time that scientists were looking for a solution to the fragility of the glass bulb of the valves.

Furthermore, to complete the puzzle around the invention of the Transistor, I must remind to the reader my experiment with the galena and its rectifying effect which physical origin was a total mystery and a mystery not only to myself but to all the scientists of that time.

Well, incredibly all those mysteries found a solution only when the Transistor came to light and its theory developed.

I want to complete here in advance the explanation for the technical reader eager to know the working principle of my galena crystal (called also lead glance or lead sulfide).

Impurities close to the surface create a kind of natural diode junctions, as mentioned in a previous chapter, and by moving a metal contact over the surface of the crystal one could intercept one of these diodes where a random combination of specific atoms that, by doping the crystal from the outside, were creating more moving electrical charges, positive or negative, as I will explain in a next chapter.

And so there is a clear parallel between the archeological galena and the Transistor which working effect was discovered on December 1947.

Both are basically semiconductor materials with point contacts standing at their surface and both are called "point contact devices", more precisely, this first Transistor got the name of "point contact Transistor".

The reader will by sure remember my difficulties in trying to move the metal contact on my galena and as well as the galena also the first point contact Transistor had that same precariousness.

This is the reason why the same scientists in the following 1948, and particularly William Shockley, managed to create a more stable Transistor called then "junction Transistor", in which the wires were welded to the crystal. This more modern Transistor was as stable as my OA70 junction diode seen on my story.

Back to my story, I was quite grown up and I started my first course at the Polytechnic of Milan. I have been reading everything I could find on my preferred subject; I found even the the text of the original patent on the first Transistor, patent issued in the United States the February 1948. I still own my first copy and I do not read in it any explanation of the physical phenomenon at the base of the Transistor.

Said the above I has been persuaded that the first Transistor, the one of Dec 1947, was not an invention but more a discovery almost by chance, a bit like my galena of which no one knew the theory.

I do believe that this argument has some historical interest and it demonstrate how some unexpected discovery presents itself and disrupts the whole future of mankind, sometimes positively and sometimes negatively. In our case by sure positively.

It is not argument of this book to discuss about the philosophical concepts around discoveries and inventions, but basically we can say that an invention is just the output of our mind while a discovery is the result of the perseverance of trying and trying again.

For what concerns our Transistor my today's conclusion, after all my readings, is that the Transistor is in between a discovery and an invention.

Invention, because since 1930 scientists had the idea to obtain electrical amplification in a solid but until the discovery of 1947 no actual result was achieved.

Discovery, because what Bardeen and Brittain were searching the Dec 1947 was not that of building a solid state amplifier but to measure surface phenomenon in a germanium crystal that the physic theory at that time did not know how to explain.

They were trying and retrying by moving the point contacts over the surface of the crystal to measure the various electrical parameters and finally their contacts found where practically did exist a natural Transistor on the surface ... do you remember the natural diode in my Galena? They discovered the amplification effect in a solid material ... seventeen years after the idea was patented!

I found the theory of the Transistor by reading William Shockley's writings who understood these phenomena and has been the first to formalize its theory. In the year 1948 Shockley invented the "Junction Transistor" and we can affirm that this was a true invention, the result of pure theoretical speculations.

Shockley published in a famous book his theory entitled "Electrons and holes in semiconductors", Van Nostrand Company in 1950.

This scientific text illustrated formally everything we are talking about and it is still the basis of all the studies that came after.

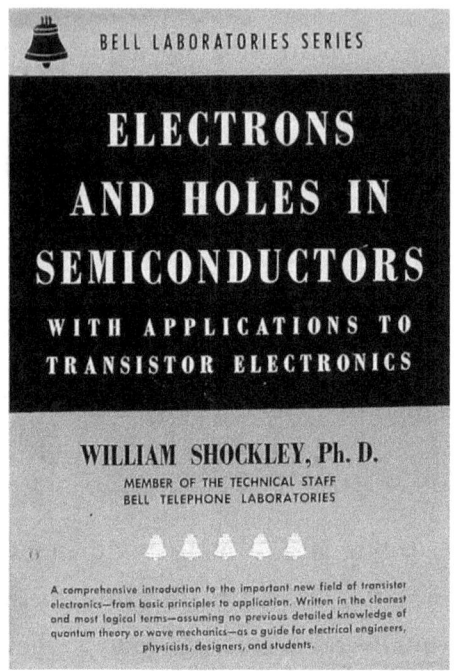

Bell Laboratories 1950. The first theoretical text on the phenomena of positive and negative charges in a semiconductor. It constitutes the scientific basis of all solid state electronics till today

While student I had the opportunity to consult it and I can't deny that I had great difficulties to understand it, especially for the high mathematics that the author uses in his book

A good book then, but only suitable for scientists with a good foundation of physics and mathematics (those who wish to consult it can access it by clicking this link: http://bit.ly/2utHEyU).

The Invention of the Junction Transistor

Even though It is not the purpose of this book to enter the complicated and tedious analyzes of the solid-state physics of which the Transistor is part, I'll try to pass to my reader the synthesis of what kept me awake for long nights during my university studies.

I should confess that only after many years since the beginning of my electronic adventure I have managed to answer to my question marks, including the already seen mystery of the galena.

At the beginning of my adventure, being childish, my curiosity was not that of understanding the "physics" of what I was building, but growing up the mysterious miracle that by sticking with a metal point the little sparkling crystal, occasionally, a sound was jumping out, forced me to look for a scientific explanation.

My naive feeling at the beginning was that the crystal itself could be somehow magic and the heart of some fantastic process, no one knew anything about and no one could explain to me.

I should say that I was really happy to hear my third grandson asking me "what is there inside my iPhone that makes it to work", showing a nice curiosity that might be an incentive also for him, like mine at about his age.

Unfortunately today the spread and easy availability of ready to run electronic devices, million times more complex of my crystal radio does not push the today young person to spend time to put together something, piece by piece, as I did and then to figure out how it works. I really hope that despite this limit the curiosity and the intimate need to research answers is surviving.

Back to the physic of these newly created solid state devices I discovered that 1948 has been a very prolific year and that there has been quite a race between Dr. Shockley and both Bardeen and Brittain to invent more.

Shockley was by sure a brilliant mind and a great physician and he immediately understood and theorized the surface phenomenon on the germanium crystal; practically he understood how that natural Transistor was made and this allowed him to replicate that phenomena but with a complete different structure: the already seen Junction Transistor!

The Junction Transistor designed and tested by Shockley didn't need any point contact: the wires were bonded to a specially prepared piece of germanium crystal. Shockley thus eliminated the fragility of the contacts of the primitive Transistor and paved the way for its industrial manufacture.

Shockley had to be a genius of solid state physics, so much so that I remember to have read somewhere a statement about him by his disciple named Robert Noyce, the Noyce that would become famous as founder of Fairchild Semiconductor and then of Intel. His sentence was saying more or less: "Shockley is a peculiar genius, he sees the electrons with his eyes!"

Thus we e can say that the Junction Transistor has been "invented" by William Shockley's mind!

He thought that the amplifying process could be achieved in a single semiconductor crystal, said monolithic crystal, artificially prepared following certain rules he described mathematically.

Shockley's equations opened the door to other scientists that so could design other solid state devices.

The following years have seen an incredible amount of new devices till the today semiconductor memories and microprocessors.

Thus, not only he eliminated the fragile contact pins but made the Transistor much more reproducible kicking off an industrial mass production that would continue until today.

Among other things, the findings, were also the basis of the phenomena in my primordial galena and now, many years later, finally I understood the mystery and any magic disappeared!

Even in the first Transistor those areas were created in a natural way, while in the new Shockley's Transistor they were produced artificially.

I suggest to the expert readers wishing to learn more about this fascinating subject to have a look at the numerous specialized texts and, if you are a scientist conversant with high mathematics and quantum physics, the book "Electron and Holes in Semiconductors" that I mentioned in the previous chapter, should be carefully studied.

To understand the semiconductors in a scientific way, I had to resort to a lot of mathematics and part of the quantum mechanics that is at the base of the description of all the phenomena occurring in a semiconductor.

In this book of mine, written for young readers, I will later on explain in a plain way the physics of our Transistor by remembering an Einstein's sentence saying that "any scientific argument can be described in a way that is understandable by everybody" ... and in our case with no mathematics and no physics.

There is another important book that I would like the expert reader to know; its content is at university level and teaches how to design electronic circuits based on Transistors.

I report here the cover scanned from my original that I found rummaging in my library among old books.

Opening it I found all my pencil notes written on the margins of the book, many of which notes, rereading, I understand how I found that book difficult too.

Despite being a very old book I verified that it is still on sale so I conclude that it must be a milestone for technicians wishing to learn this matter from an authoritative author.

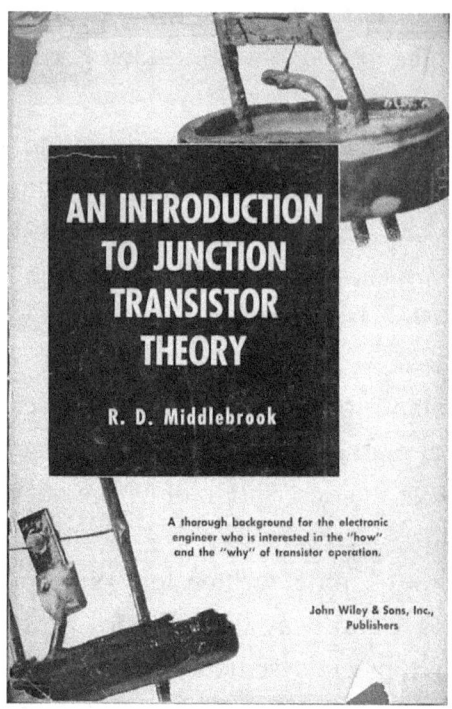

1958. An Introduction to Junction Transistor Theory, by R. D. Middlebrook, edited by John Wiley& Sons. My basic book during my university courses

Electric Current: what is it?

Before to start explaining how the diodes and the Transistors work, subject of the next chapter, we must know what electric current is and how electric current works and this chapter is dedicated to this argument.

We'll get into some details concerning the atomic physics, but you will see that these concepts are not as unreachable as one may think and you, at the end, shall know the phenomena that are the basis of most of the electronic components that are in use today.

To Summarize, we have seen how hertzian waves were reaching the radio antenna with their knights, audio signals, we have discovered how the diode was able to stop these hertzian waves and leave to pass just the audio signal: technically the diode was "rectifying" the high frequency of the carrier, leaving a signal at much lower frequency, precisely an audio frequency.

Then, the Transistor came into the scene, took the weak signal from the diode and it made the signal stronger by amplifying it. In the next chapters we will go deep into the diode and into the Transistor to understand what is happening there.

But to understand the internal physics of semiconductor devices we need to achieve some knowledge of the mechanism that relate to the electrical current, electrons, crystals and atoms.

Before starting to write what follows and to be sure that all my readers, including very young students, could understand what I am saying, I decided to check at what level of knowledge were the skills of my nephew and also that of my wife, notoriously far from technical issues.

So I asked if they had ever heard something about the atom, electrons and electricity.

I got a totally negative response from my nephew, but my surprise has been that my wife, so often at my side in business and sharing many technical conversations, was at the same knowledge level as her grandchild (… I'll hide this statement from her eyes when she, more conversant in English language than myself, will revise my text).

Being them too part of my target, I saw the need to spend a few words as simple as possible about the argument.

To start I thought to invent a new fun character like I did with the Semimetallic Monsterling and I started saying to my grandson: "electrons are tiny animals, called Electronlings, animals with some very unique features".

They are not like the Monsterlings that we have been talking about at the begining and that in millions live inside your iPhone, these Electronlings are much smaller and each Monsterling is composed by billion of billions of them.

I see that my new fun character intrigues him and I believe I took the right path to solicit his curiosity and in fact he has immediately asked: "but what they do and how can they stay so many inside a so small space?

I thought I could make a step forward and without frightening him I could start by introducing the concept of the atom.

I tell him that these Electronlings are a fundamental component of nature, and this component is called "atom" but, I add "do not think about the atomic bomb with which you play on your video games, I am talking about a very important matter and that has nothing to do with games or with weapons".

As far as electricity is concerned I could easily bring to him well known examples like appliances, power stations and all those object that get supplied by electrical current.

More I explain that there are two basic kind of electrical currents: the one very strong and the one weak.

Virtually all equipment that we hold in hand or on which we operate directly like computers, cell phones, cameras, etc. work with weak currents, voltages of at most of a few tens of volts and for that reason defined weak.

When we see the power lines suspended on poles that run to the sides of the roads and the cables between the high voltage pylons then we are definitely looking at "strong currents".

My grandson likes these explanations and to make even clearer what I am saying and more interesting I show him some pictures and I tell him:

"If you're with daddy in his car and if you pass in front of a power plant with huge chimneys from which smoke comes out, there large amounts of electric current are produced".

"The windmills that you saw in Holland produce strong currents too"

"The solar panels you see on the roofs of the houses and in the fields are exposed to the sun so that they can transform the Sun light into strong currents to light your home by night".

What atoms and electrons have to do with all this? Here lies the greatness of humanity who manages to wrest the electrons from the atom and to divert them on metal wires and send them up to our houses power socket!

The next chapter is dedicated to explain how atoms and electrons work for you.

Atom and its Electronlings

Let's start by saying that the electron, that here we nicknamed Electronling here, is an animal that lives in a small house called atom, better yet, in the atom he is a great wanderer and runs continuously around the center of his house, tirelessly and without ever stopping.

Electrons have a negative electric charge

He is living with many siblings, and they all revolve around the center of their house without never stopping and they have a peculiar characteristic: they bring an electrified suit loaded of negative energy with which they play as we will soon see.

What are the atoms for? Every element, every substance, the stars, the planets, the whole universe and even ourselves are formed by billions of billions of billions of atoms and every element of nature is distinguished from others for how many Electronlings has in its atoms.

For example in silicon every atom has 14 Electronlings, Germanium has 32, gold 79, oxygen 8 and all these hungry

Electronlings revolve around what is called the nucleus and together they form a little house that is called atom.

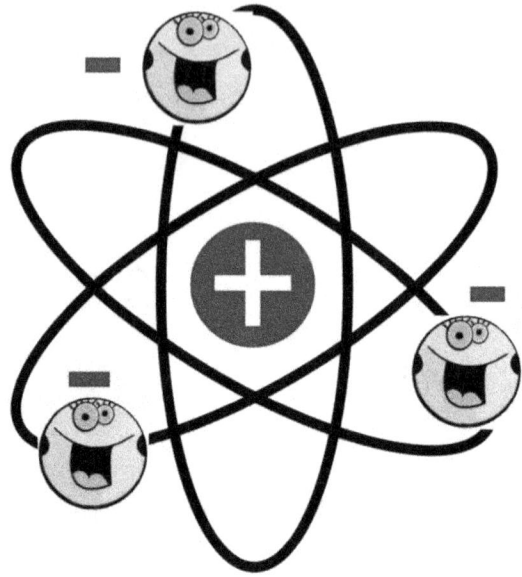

An atom is composed of a positive nucleus and electrons that turn around it

That's the atom, the basic component of all known matter and that is combined in various ways to form those beautiful structures that we call crystals and at the center of this house there is a positive charge exactly equal to the sum of the charges of all the Electronlings that turn around but of opposite sign, i.e. a positive charge.

This is true for all the elements of nature but the crystals have a peculiarity in addition: their atoms are very well organized and perfectly aligned and each atom occupies with its own house a space that the condominium, that is the crystal, assigns to each atom with precision. We will see the importance of this peculiarity of crystals when we will talk of the semiconductors.

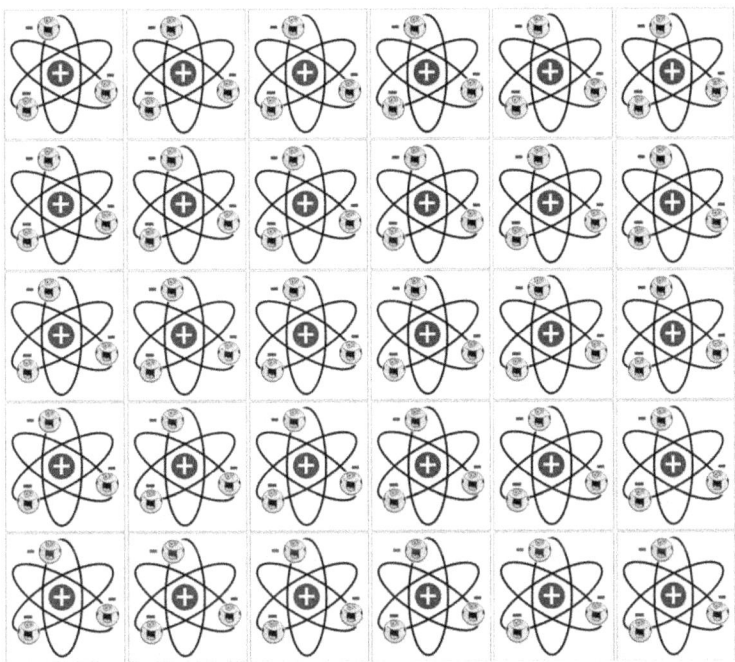

In a crystal the atoms are aligned according to a precise pattern

We now know that each element of nature is formed by many atoms, each of which has a number of negative Electronlings that run continuously around a nucleus.

The atom, as we have seen, is a bit like a playhouse for these Electronlings brothers but one strange thing happens under certain conditions.

In all respectable home, these Electronlings should stay there where the family would like, i.e. turn around inside the house and not go for a walk when they want.

In a true family if the child comes out on his own, then he is brought back home and put into punishment.

In some elements of nature happens just that: those Electronlings which runs in the outer part of the house are also the most rebellious and often they come out without asking permission. If they find other companions that have come out from nearby houses they form a team that runs away.

These Electronlings escaped and that formed a team that roams freely without a precise destination, they become what technicians call "free electrons".

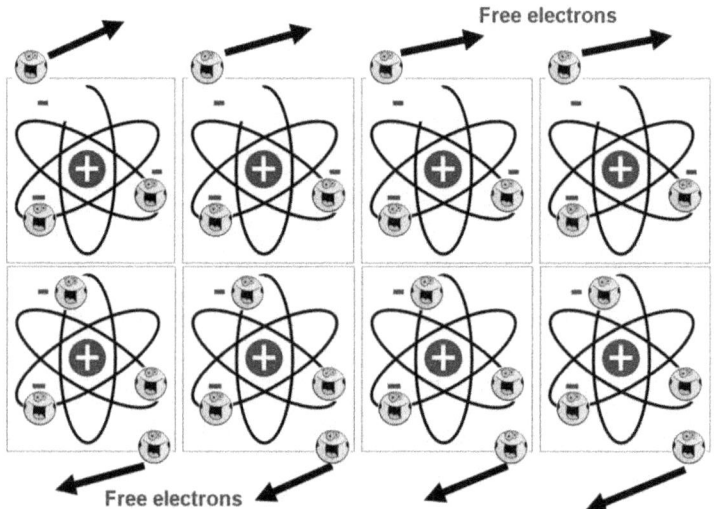

In a metal abound free electrons to move

And now comes the fun part: these squadrons of moving electrons that form precisely that thing that we call electric current can be captured by external forces, as we will see soon.

Not all materials have electrons so rebels to run away and circulate free out of their house, this happens only in certain elements that are for this reason called conductors.

All conductors like copper, silver, iron and other metals have the peculiarity of having inside an ocean of electrons that roam free. These substances, that have a large amount of free electrons, can carry large amounts of electricity and for this we call them "conductors".

Now we can talk about what happens to a metal containing many electrons that have a negative charge if we connect a battery as in the following picture: those rebels Electronlings (the electrons) are attracted from the positive side of the battery and form a flux of moving electrons; this flux is what we call "electric current".

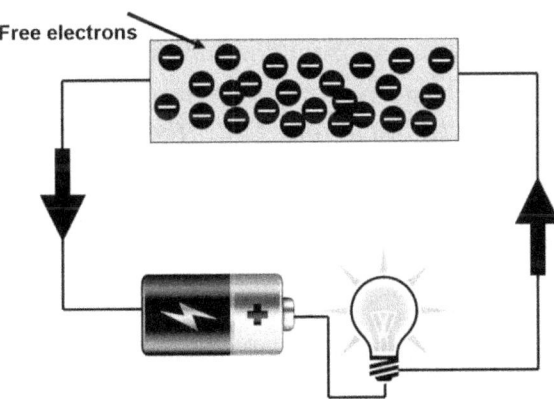

A flux of electrons in a conductor is what we call "electric current"

Even if we reverse the battery the same current reverse its direction carried by the moving electrons.

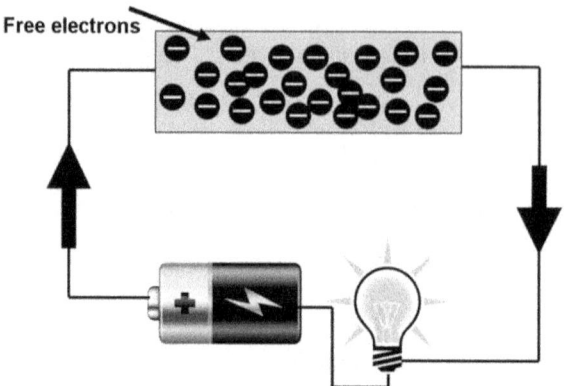

By inverting the battery the current flows perfectly in the opposite direction

Remember this fact, because soon we will need it: in a conductor the electric current can either go back and forth.

It must be outlined that by convention the electric current flows

> *Important: in the 19th century it was decided the conventional direction of electrical current to be from the positive pole to the negative pole. Then, the electrons move in the opposite direction than the conventional electrical current direction.*

from the positive to the negative pole of the battery and this is the other way around to the real movement of electrons.

What today it appears as a strange convention, actually derives from a choice made when humans still did not know what electricity was and even humans did not know about the existence of electrons.

Let's move on to another situation. If we connect a battery to an insulating material as shown below and as we know the insulator has no free electrons that can move, no electric current can be present here.

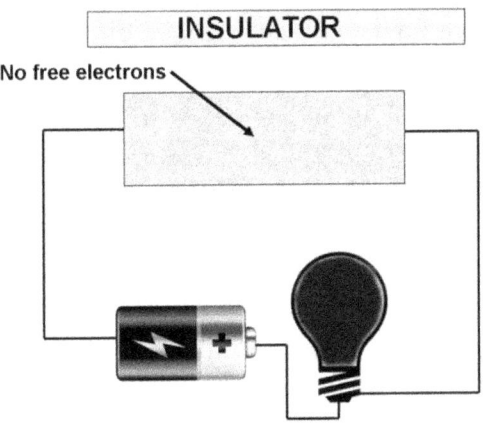

*In an insulator a battery doesn't find electrons
to push and no current can flow*

If we overturn the battery the situation does not change: the insulator will not allow any current to flow as in the previous example.

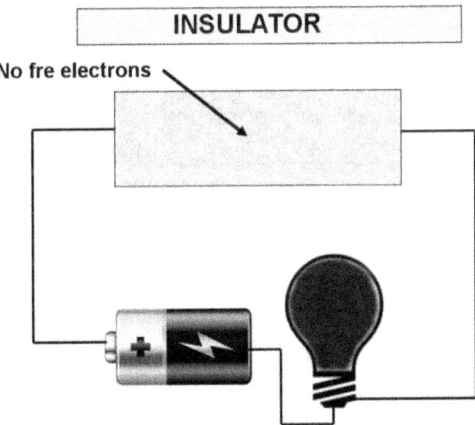

By reversing the battery no current flows

Therefore an insulator never allows an electric current to circulate and this is the reason why we use these materials to isolate us not to run the risk of being electrocuted.

In fact, if we touch an electric cable with a wooden stick, which is insulating, we do not run any danger, but if we touch it with an iron bar, which is a conductor, we risk of being electrocuted.

In conclusion we learned how the electric current is formed in a metal, i.e. applying an electrical source, battery or current socket, and how those free electrons that are inside it begin to flow and how they end up lighting a bulb, operating a radio or an iphone.

Those substances that have this property, i.e. many free electrons, I repeat, are good conductors of the electric current.

By contrast there are other substances where electrons are kept very tight in their houses, namely in their atoms, and they are forbidden to go walking around freely, to form a team and do what their fellows in conductive substances do: they are poor prisoners.

They are, for example, ceramics, wood, marble, plastics and many more that we use to isolate ourselves from the power supplies and also from lightning.

At this point we have splitted the substances into two large categories from the electrical performance point of view: conductors and insulators.

At the beginning of this book we have being talking about a substance that we called "semiconductor" and the careful reader can now imagine that we are at the right point to call it back here and to compare it with what we've seen in this chapter.

It is quite obvious, also from the name, that a semiconductor should be something in between a conductor and an insulator.

Semiconductors in fact are half conductors and half insulators: that is, even in these substances there are free electrons that go around by themselves after having left their atoms, but their atoms are very careful and they keep the most inside their houses and the free electrons that can escape are just a few.

We will see in the next important chapter how semiconductors behave and why their use is key to the electronic world.

Semiconductor s: how they work

We saw that a conductor is able to conduct electricity well because it has a huge amount of free electrons while an insulator has virtually no free electrons and therefore cannot conduct any electricity.

In a pure semiconductor, instead, circulate few free electrons and the electric current that these electrons manage to carry is infinitely smaller than that of a conductor like copper, but greater than that of an insulating material such as ceramic.

Anyhow our semiconductors have another interesting feature; it is true that they have few free electrons to move than metals but we can artificially add more electrons from outside and therein lies the origin of the possibility to create semiconductors with the desired characteristics of electric conductibility.

How is that possible? Simple, adding atoms each of which has an electron ready to leave and to turn free once it has been inserted into the semiconductor.

By inserting atoms with an electron in addition, the latter becomes free to move

What does this mean? That we can act, for example, on the silicon crystal and adding, with appropriate techniques, some particular atom and to get a doubling, tripling, and even beyond, effect of its free electrons.

We can compare this process to the melting of salt in the water to make it more salty: depending on the quality and quantity of salt added to the water we get a water with different characteristics.

Well, with semiconductors is something like that, with the difference that it is easy to dissolve salt in the water, while in our case we have to tuck it inside a well-solid crystal and here lies all the difficulty.

So we come to the fundamental property that allowed the industry to manipulate these semiconductor substances, and then to change at will the ability to conduct electricity, meaning to bring it to the values that the scientists calculate so that the effect diode and Transistor can be achieved.

Let's see how it happens in reality and remember that a crystal consists of atoms all well organized and aligned, many small houses in a well-ordered condominium where every house has a place exactly assigned.

Now imagine sticking in this condo a cottage a bit bigger, say that instead of having only three electrons have four; well, if it wishes to stay in this condo there is little to do, it must lose an electron and guess what this electron left alone will do: this electron becomes just a free electron to move inside the crystal and then to become an element to conduct electrical current.

We have therefore found a way to increase the free electrons of our semiconductor crystal and since the electrons are negative, this crystal the technicians call it "N".

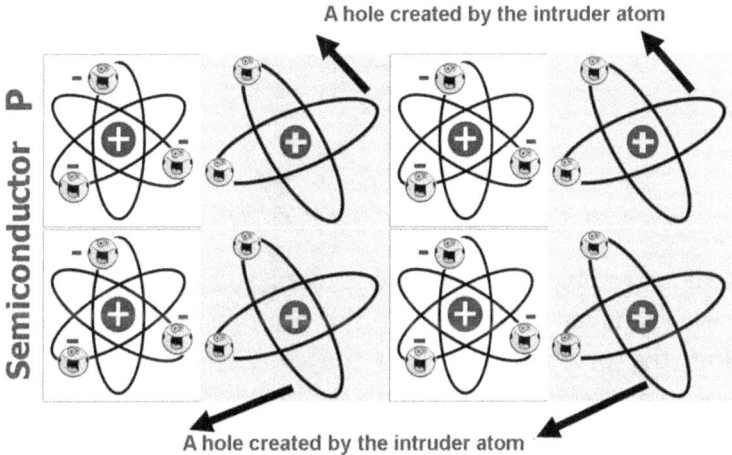

The collection of excess electrons create a N semiconductor with free electrons

But the story does not end here: once you are sure that you can force a fat atom to squeeze into a smaller space and forced him to slim down leaving one of his pieces, that is, the electron, free to go, it was decided to do even another thing: follow me.

Imagine to have a smaller atom, with only two electrons, and want to force into our semiconductor crystal and let him occupy the space of a house, what happens?

At the end he has a hole, that is, an electron less than its neighbors, and might happen that he takes away an electron from the nearest house.

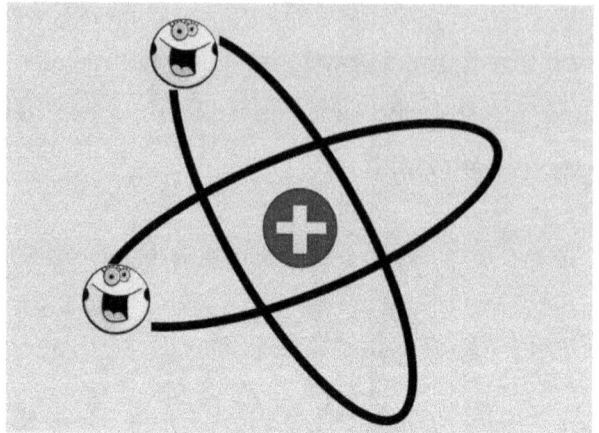

By inserting atoms with an electron less a free hole to move is created

It seems incredible but we have created a crystal in which atoms steal electrons and because in the atom where the electron is missing, the positive charge in the center is prevailing, all this equates to an electric current in the opposite direction to that of the electrons, that is, they move virtually the positive charges and these charges are as many as the thin atoms that we have inserted into the crystal.

Take heed to this important fact discovered in the last century and that Shockley, one of the inventors of the Transistor.

Shockley has used this property to design the junction Transistor creating a crystal that can conduct electricity despite not having free electrons.

A Crystal where "lack of electrons", i.e. holes, can move as a free electron but in the opposite direction, just like a positive charge should do.

For this reason, a semiconductor that conducts in this way is called "P" for positive, i.e. with apparently positive charge carriers.

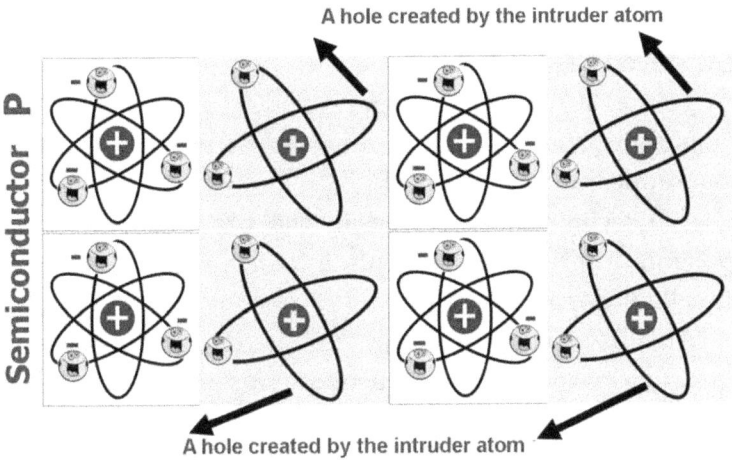

All the excess holes creates a semiconductor P with freely moving positive charges

Do you remember the galena? That material too is a semiconductor, and though it was necessary to continuously move the metal point to find the area with the right amount of free electrons it was working as a natural diode.

With modern means and scientific knowledge we are able now to determine where and how that feature has to be: on this basis are manufactured diodes, Transistors and solid-state electronic components, including microprocessors.

This peculiarity of having few free electrons and to be able to increase or decrease them at will is essential to the fabrication not only of diodes and Transistors but of all modern electronic components.

To tuck into the Crystal the substances to change its structure cannot be obtained simply like to salt water where simply by sprinkling salt in water we get what we want; in our case it is necessary to act in a way much more energetic.

The crystal should be put in a special oven filled with the substance to be inserted in the Crystal properly vaporized and at certain temperatures this substance is forced into the crystal structure and, depending on the type of substance, the semiconductor becomes N or P.

For simplicity in our examples we have used atoms with two, three and four electrons, but in reality the germanium has 32 electrons and the silicon has 14 electrons.

To create new free electrons and free holes must be used elements that have an electron more or un electron less than the original semiconductor.

The technicians involved in the production of electronic components insert these elements into the furnace for a process that is called "doping".

At this point we know how we can produce P-type semiconductors and N-type semiconductors, and we can reveal the mystery of the diode and its particular way of conducting electricity.

You will remember that the electric current is nothing more than the flow of free electrons and that the diode has the particular characteristic to slide the current in a sense and prevent it from flowing in the opposite direction: what we learned till now will bring us the solution in the next chapter.

We summarize these two key points:

1 – It has been discovered that in a semiconductor crystal the electric current can be transported even by free holes, i.e. by atoms where an electron is missing, and thus prevailing the positive charge of the atom in the center, these pseudo mobile charges are called P.

2 – We can artificially create semiconductor with free electrons, called semiconductors N and semiconductors with free holes, called semiconductor P.

How the Diode works

Clarified how electrical charges are carried out in a semiconductor crystal let see what will happen if we adjoin a N-type semiconductor crystal with a P-type crystal, i.e. a single crystal that have respectively negative free electrons at one side and positive free holes at the other side.

In this way we form what is called a PN junction, in reality it is not formed by gluing two separate pieces, but spreading in a single crystal on one side and the other those elements necessary for generating free electrons and holes at each side.

The crystal then is one piece ("monolithic") and has no physical cut in the middle and so the charges can move freely from one side to the other.

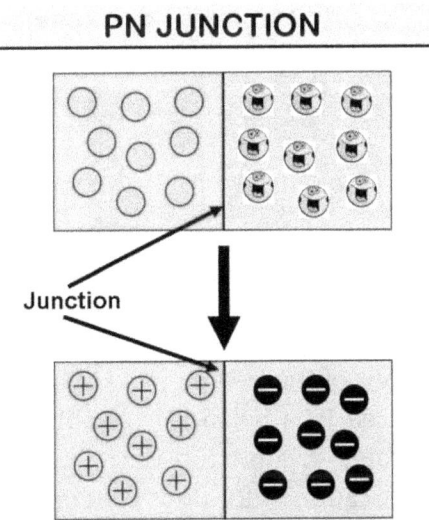

By joining two zones, one P and one N, a PN junction is formed

Let's clarify what's going on around the junction where the type of charges changes and where the opposite charges overlook at each other.

The charges close to the junction tend to attract the one of opposite signs that lie on the other side and then move through the junction, but there they encounter small animals that immediately begin to quarrel and they end up annihilating each other.

The opposite sign charges are attracted and annihilated around the junction

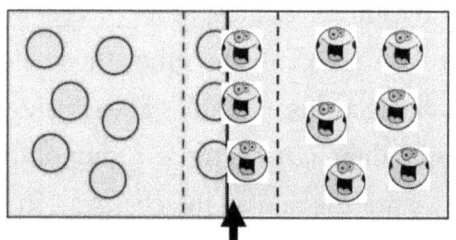

Here the charges beat each other

... and eliminate each other

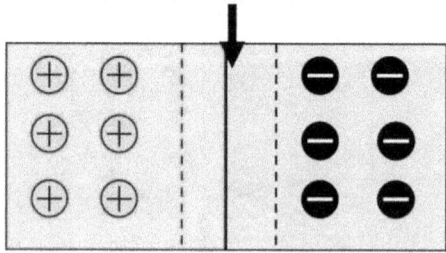

At the end it is formed an area in equilibrium with no mobile charges

But once the electrons of the N-type will have occupied the seats of the lodges in the area P   a balance is established and everything stops forming an area around the junction that technicians call "depletion layer" because it does not contain neither positive nor negative charges.

This area has an important role in everything that happens in the diode and in the Transistor that we will see later on.

For the time being we should only remember that the functioning of all solid state devices depends from this junction that can be crossed by electric charges, once created.

So, as soon as the crystal with its P-N junction is formed in production, the various electrical charges, P and N, are stabilized and each one remains in the zone that belongs to each one and they will not move till an external event intervenes to change its equilibrium.

Now let's see what happens if we disturb them by connecting to our crystal a battery as in the following picture:

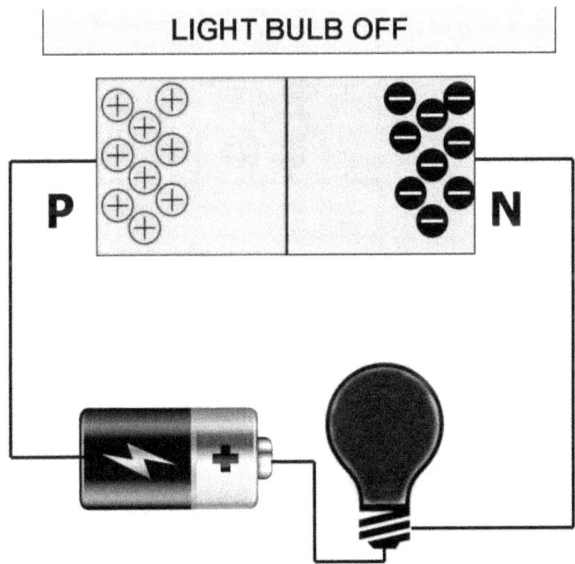

By connecting a battery in reverse mode, more charges are removed from this junction and no electric current can flow through

If we connect the negative pole of the battery to P and the negative pole to N the various charges are attracted by the respective poles, leaving the junction and creating a larger depletion layer with no mobile charges in it.

As we remember, a substance which has no moving electric charges is practically an insulator and also this diode polarized in this way behaves as an insulator and does not allow electric charges to move in the circuit: the lamp will remain off.

It is said that if we polarize the diode in the reverse mode, the positive pole to N and the negative to P, the diode acts as an insulator.

Let us now look at what happens if the battery is inverted, as in the picture below.

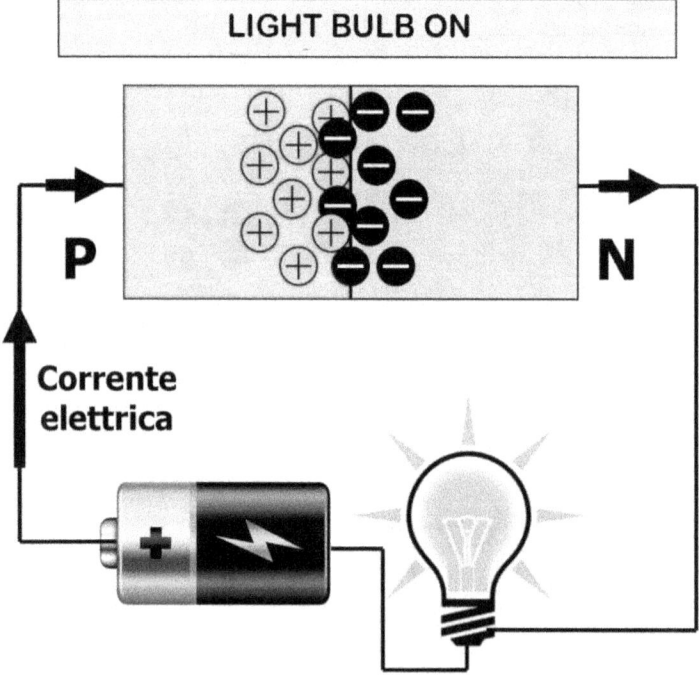

By connecting directly the battery, charges are pushed beyond the junction and the electric current can flow continuously supplied by the battery

Now the positive pole is connected to P and the negative pole to N: but the same charges repel each other so the free charges are pushed throughout the junction and the battery will attract charges coming from the opposite zone creating a continuous movement that the battery can support.

In other words our diode behaves like a conductor and the light bulb will turn on and remain lit until the battery will continue to supply energy.

We now know the wonderful properties of this crystalline semiconductor that has been properly doped by a manufacturer: it can behave as an insulator and as a conductor according to how it is polarized by an external source of electric current.

Every time we want that an electric current moves in one direction only and not the other way around, we have to insert a diode in our electrical circuit, silicon or germanium, and we will get the rectifying effect that we know well by now.

If we go back to our radio circuit we can now understand how the diode made possible to hear the audio signal by extracting only the audio frequency. The electromagnetic signal, or Hertzian wave, at the antenna fluctuates between negative and positive values and the diode cut the wave leaving only one part of it that performs like an audio signal.

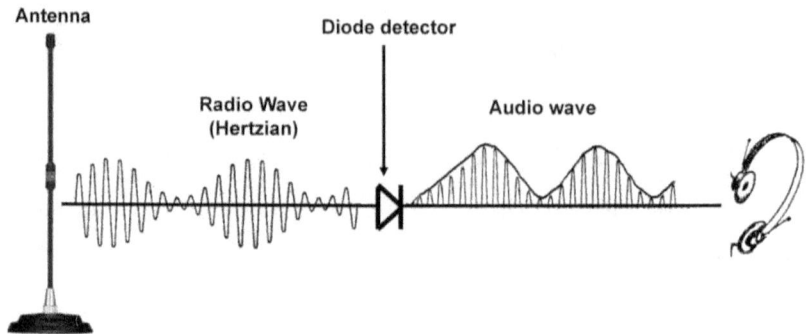

The diode lets pass the positive part of the signal while blocking the negative part by creating the "detector" effect

We know finally how the P-N junction works, we learned what electricity is, we saw how you could artificially alter the electrical characteristics of a semiconductor crystal and now we are ready to face a much more complex device than a diode: our hero, Mr. Transistor.

How the Transistor works

We will treat here the bijunction Transistor that is formed by a semiconductor crystal, germanium or silicon, where have been created three zones, electrically different and, according to the way in which these zones have been doped, are called Transistors NPN or Transistor PNP.

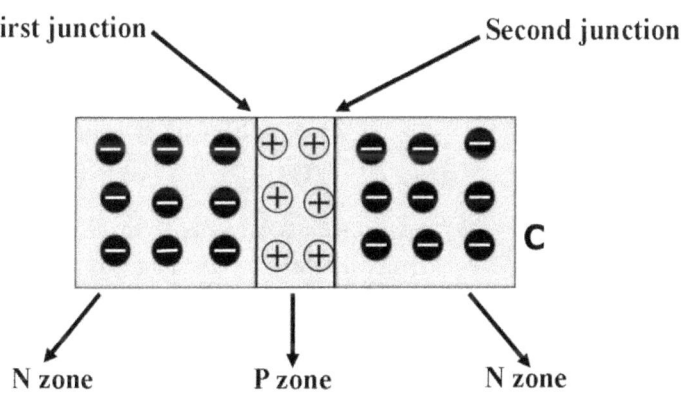

The junction Transistor is formed by three alternately charged zones and a very thin central one

This is the first type of Transistor really industrialized and it is precisely what Shockley has conceived in 1948 to overcome the technical difficulties of using the first point contact Transistor.

After the invention of the junction Transistor, many other types have been created, each with its own characteristics and advantages, but the founder of all is definitely this Transistor.

In this chapter we will learn how it works, all the others are derived from it and we we'll see them synthetically in the following chapters.

Imagine to attach to an NP diode a second PN diode and to form an NPN structure as in the following picture.

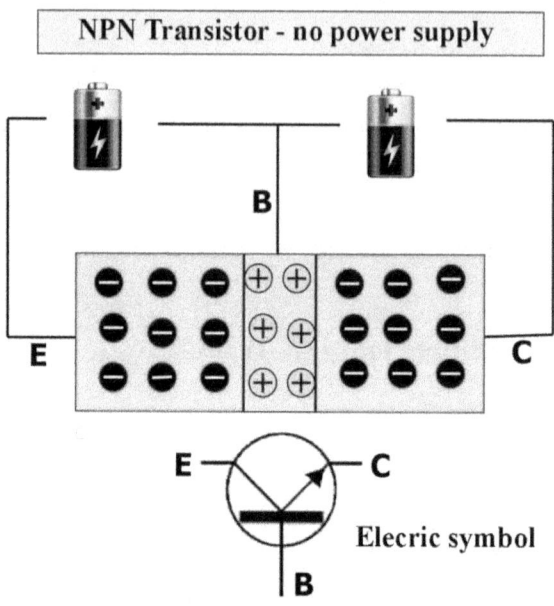

The Transistor is formed by two junctions and needs to be supplied by two batteries. In the figure, the batteries are disconnected and the three contacts are called E = emitter, B = Base, C = Collector

Here though, to operate the Transistor, as the technicians want, and not just to rectify an electric current, but to amplify a signal, it was decided to make the P zone in the middle very thin and with few free charges, though much less than in the other two N zones.

Be careful: the three zones in the real Transistor are part of a single semiconductor crystal where the free charges have been doped into it in a oven at high temperature.

To understand how this device works we must remember what we learned in the previous chapter dedicate to the diode.

If we connect a battery at a junction in a direct mode (forwarded biased), i. e. the positive pole of the battery to the P zone and the negative pole to the N zone, the junction behaves as a conductor, while if the battery is reversed (reversed bias) the diode behaves like an insulator.

But in our Transistor we have two junctions and the free charges inside do not move if the two batteries, as in the previous picture, are not connected. What will happen if we connect the batteries as in the next picure?

Here the miracle comes! If we connect the first battery in a forward way and the second in a reverse way, the first junction acts as a conductor and the second as an insulator.

All the internal mobile charges are somehow stressed by both batteries, apparently one in opposition to the other.

It happens that the first battery is pushing violently the mobile charges, which in this case are electrons, throughout and beyond the first junction, i.e. the P zone in the middle.

But this P zone is very thin and has just few positive charges, so jast few negative charges from the left N zone will be annihilated and the most of them will reach the third zone on the right

NPN Transistor - Power supply ON

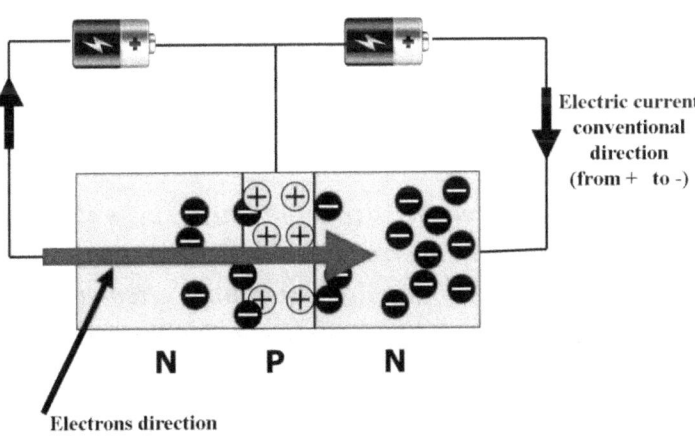

The first NP junction is forward biased, the second PN is in reverse mode and being the P area very thin, the electrons pass through it

What we will see now is the key point to understand the amplification in a solid material, i. e. in a semiconductor.

What did Shockley thought? If in the zone in the middle, a P zone in our case, we make sure that there are just few positive charges and we make this zone very thin, the negative charges that arrive there at high speed pushed by the battery, not only will find just a few charges to recombine with, but they will reach the second junction and will enter the third N zone.

But the third N zone is powered by the positive pole of the second battery and so they are attracted by that pole and an electric current will start to flow powered by the battery.

Thus we have a stream of electric current formed by electrons that continuously passes from the first zone to the third powered by the two batteries as if a conductor existed between the two poles of the battery.

But this strange conductor has something in the middle: a P zone!

Now it is obvious that this zone must have some effect on the charges which run from the first to the third zone. In fact if we introduce other charges here, P or N does not matter, the crossing electric current will be modified.

This is exactly how the junction Transistor performs its amplification effect.

> *The Transistor effect consists in modulating, by a weak signal applied on terminal B, a stronger electric current through the first forward polarized junction and that, after crossing the second reverse polarized junction, it reaches the terminal C as an amplified signal.*

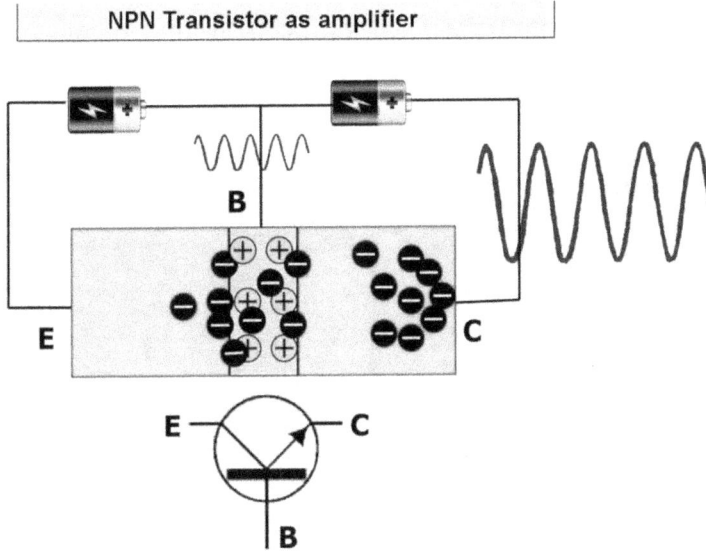

A weak signal on B appears amplified in C and here it is supported by the second battery that is working on a diode in reverse mode. Such a diode has a much higher internal resistance than the one between B and E and so we have the same current over an higher resistance and this is an amplification effect.

The above figure shows the electrical symbol of the junction Transistor and its terminals names are:

E = Emitter   C = Collector   B = Base

We can now summarize by saying that in a junction Transistor, if the first junction is forward biased and the second junction is reverse biased, a signal applied to the zone in the middle, will appear at the terminal C amplified.

We have reached our goal to understand how our solid state device works as an amplifier in its basic configuration.

As such he became a fundamental constituent of all modern electronic circuits, including the iPhone that my grandson was holding in his hand.

To conclude our chapter, here below you can see the actual section of a modern silicon Transistor produced by the so called "planar process", the same process used today for the production of all integrated circuits, processors, memories, etc.

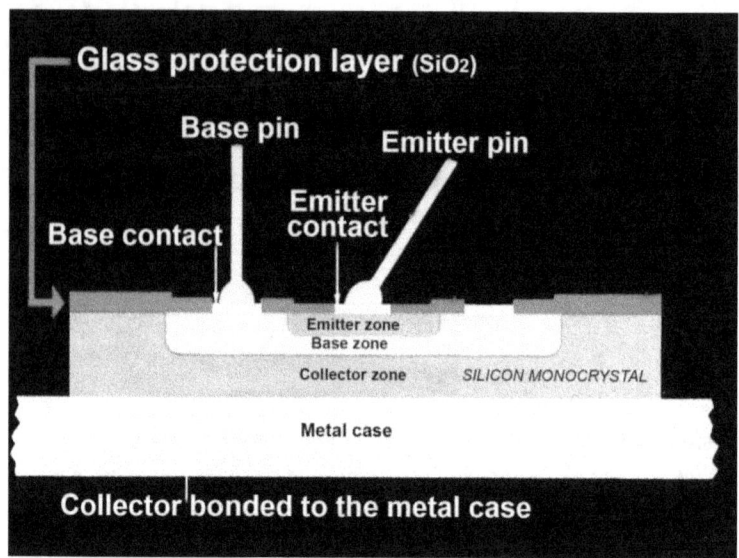

Section of a power planar junction Transistor

Our conversation should now proceed with some notes on the most important siblings of the junction Transistor, many of which are today a preponderant part of both civil and military electronics.

We promised that this tale should be entertaining and instructive at the same time so now let's interrupt the technical part for a while and let's talk about some curious facts to which I didn't give much importance many years ago.

At a distance of time they have gained a great value, to begin with my first visit at the small Intel factory at Mountain View, at a time when the today Silicon Valley was still called "South San Francisco Bay area".

My early Days at Intel

It is exactly on March 15, 1970 when, first time in California, I arrived invited by the just founded Intel Corporation.

I visited this small factory at 365 Middlefield Rd, in Mountain View, a few minutes' drive from the famous Stanford University of Palo Alto.

To receive me there was the number three of the organization, namely Bob Graham, the then marketing manager, and we have been photographed in front of the factory.

It was March 1970 when I visited for the first time the newborn Intel in Mountain View. Waiting in front of the entrance there was Bob Graham, marketing manager

On that occasion I was fortunate to meet the founders, Bob Noyce and Gordon Moore, and to shake their hands at a time still at the beginning of their new adventure that would become one of the most important company in the world.

I met also Andy Grove and Ted Hoff and I could see their first laboratory and the first memory production chain yielding just a couple of products.

For what concerns Dr. Federico Faggin, the famous co-inventor of Intel microprocessors, he was not there yet, but I had the privilege to see him drawing the first microprocessor chip, the i4004, the following year.

Before leaving Intel to return to Italy I photographed from my rented car the front of this early factory and I am sure that this picture is quite unique even at the Intel Museum.

Before leaving Intel to return to Italy I photographed from my rented car the front side of this first Intel factory and I am sure this picture does not exist even in the Intel Museum.

The reason for my visit was a training to deal with the few products that Intel was delivering at that time and on that occasion I got the first technical documents on Intel memories like the i3101 and the first 1024 bit dynamic memory labeled i1103.

I was also informed about the selling prices and the first Intel price list signed by Bob Graham has been handed over to me, original that still I own.

 INTEL CORPORATION 365 Middlefield Road, Mountain View, California 94040 · (415) 969-1670

Dear Sir:

Thank you for your interest in Intel. We have enclosed the product informatin you requested.

The Intel 3101 is a 64 bit random access memory. Its high speed makes it ideal in scratch pad applications.

Intel's 1101 is the first fully decoded 256 bit static MOS RAM. It utilizes a new silicon gate technology which allows much greater component density and circuit performance.

The price breakdown for these circuits is as follows:

1101	Prices
1 - 9 units	$150.00 each
10 - 24 units	$110.00 each
25 - 99 units	$ 80.00 each
100 - 249 units	$ 65.00 each
250 - 499 units	$ 55.00 each

3101	Prices
1 - 9 units	$ 99.50 each
10 - 24 units	$ 74.00 each
25 - 99 units	$ 53.00 each
100 - 249 units	$ 43.00 each
250 - 499 units	$ 38.50 each

Intel circuits are immediately available from your local Cramer Electronics or Hamilton Electro Sales Distributor.

Very truly yours,

Robert F. Graham
Director of Marketing

The first Intel price list issued by Bob Graham indicated prices per bit millions times more expensive than today

We must notice that a single 64-bit memory type 3101 was costing 99.50 US$, that is more than a buck and a half per bit!

The average cost of a today memory would be millionths of a dollar; in less than fifty years the price has shrunk by over million times, a nice leap! And this is due to the Moore's law, that Gordon Moore who founded Intel, a very important law for his technological implications and that we will see in a dedicated chapter at the end of this book.

The i3101was the first memory produced by Intel and the first product that I had the honor and the pleasure to deliver the December 1969 to Ime Spa, an Italian client in Rome, who was manufacturing electronic calculators.

Here below the first technical document issued by Intel; on the back it shows July 1969 as the print date. Surely this is the first official document ever published by Intel.

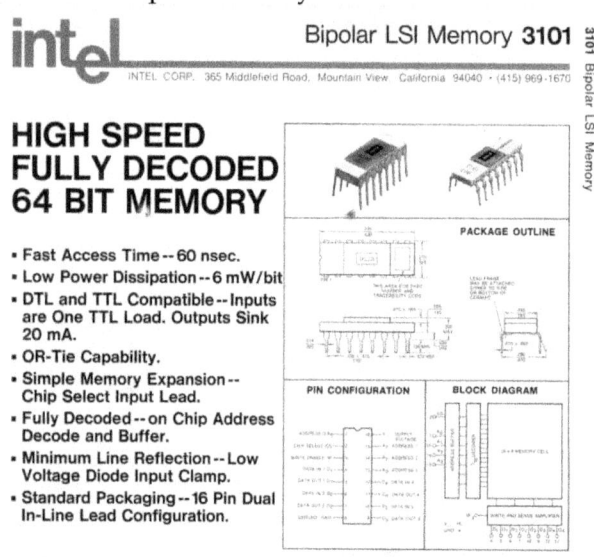

64 bit bipolar memory. First official document issued by Intel exactly 11 months after its foundation and first product delivered to the market.

As for the cute Bob Graham, who kindly escorted me for all the three days of my visit, I must add that I had no other chance to see him again as he was forced to step down soon, or so I was told, for a serious quarrel with Andy Grove, at that time chief of engineering and who will become president of Intel in 1986.

Here below three pictures of the three leaders at Intel, pictures that were handed over to me as images to be inserted into any local advertisements in Italy.

| Bob Noyce | Gordon Moore | Bob Graham |

The first pictures of Intel leaders. From left Chairman, Vice president and Marketing Manager

I like to remember my second visit at Intel the following year, precisely the August 1971, when Intel had just moved from the small factory in Mountain View to a brand new factory in Santa Clara, factory that was still to be completed when the following photo was taken.

The following year Intel moved to a new plant in Santa Clara that I visited while its construction was still underway

And it was precisely on that occasion that I had the opportunity to visit the department in which it was developing the first microprocessor and to speak with Federico Faggin while painting the layout of his 4004, first microprocessor, and for many years I has been following his new ventures from Zilog to the last Foveon.

I must say that the microprocessor became the most important and most popular electronic device derived from our old friend, the Transistor. To the microprocessor we will dedicate an entire chapter being this electronic component the heart of all computer and the product from which all the Intel fortunes will come from.

Business and Convivial Meetings

After so many information, technical and historical, the reader is probably in the need of a breaks and to this end I like to insert a sympathetic chapter as interval.

During our many technical and sales meetings organized by Intel it was customary to alternate the long hours of work with convivial breaks.

And so we also do now a breaks by describing our business meetings and their convivial breaks.

It has been fun for me rereading my old agendas that I still have and I can now report some of those meetings for which I also have interesting photos.

The first sales meeting I attended was organized by Intel at the Hyatt Regency Hotel in San Francisco between the 17 and the 21 September 1973, as I read on my agenda.

On that occasion I met for the first time Ed Gelbach, Bill Davidow and Mike Markkula, the last one Sales manager at Intel.

Markkula will become the first financier of Steve Jobs and the first CEO at Apple Computer in 1976 where I had seen him again.

I believe that this was the first international Intel meeting to which distributors of various countries were invited, including Tekelek from France, Industrade from Switzerland and myself from Italy.

Intel had organized this meeting so wonderfully with moments of technical and selling courses interspersed by various performances in an unique environment.

It was my first international sales meeting, American style, in a great hotel and in front of the beautiful Bay of San Francisco. I still remember my impressions lucidly: it seemed to me to be in another planet. I has never been in such big hotels and everything seemed to me like a daydream.

The various managers of the company spoke and I have not forgotten the clear presentation of president Bob Noyce, of whose authoritative competence everyone was aware, and I took diligently a series of notes.

The following year, namely between June 9 and June 11, 1974 Tom Lawrence, the new International marketing manager, organized on the island of Palma de Mallorca in Spain another great sales meeting addressed only to the European sales force.

In Palma de Mallorca several U.S. Intel managers assisted as instructors and among them I can remember Gordon Moore, Ed Gelbach, Mike Markkula, Bill Davidow and Phil Spiegel.

My agenda reports an interesting graph on the future of the microprocessor presented by Bill Davidow in which appears for the first time the idea of the future microprocessor generations and that impressed me for its novelty and I copied: it was the basis of what we see today! Here below that chart as shown by Bill and scanned from my agenda:

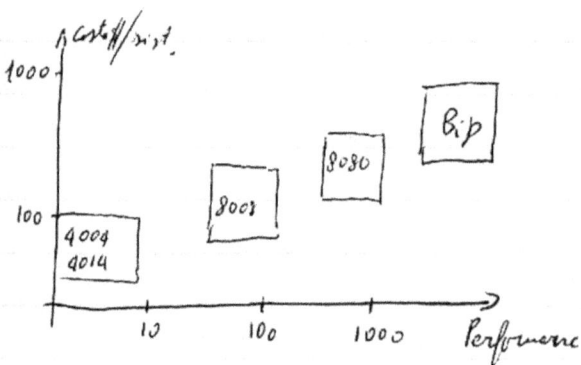

Jun 9, 1974. Bill Davidow's first microprocessor plan

In the same presentation my agenda reports a Bill statement that sound: "I foresee 25,000 applications per year for the microprocessor 4004 and I can announce the indicative price of the new 8-bit microprocessor 8008 equal to $150 for 500 pieces". At least this is what my note states exactly. Interesting true?

I found just two pictures of this meeting. The first in which I took the nice flamenco dancers.

Palma De Maillorca. June 10, 1984. Intel sales meeting. Flamenco dance and back head of Intel managers

The show was offered to us by Intel and I took it from behind the heads of some Intel managers ... of whom I was unable to trace back the names.

Judging from this photo one may think that I was interested just to the dancers. But rest assured, I was accompanied by my wife Eva that I photographed, see next photo, mainly to take the Microma watch at his wrist, a watch made by Intel.

This watch is part of the first quartz watches in the world invented by Intel and all of us were sure we might find a big market success.

In reality, things did not turn out very well for the Microma watches. We did not realize that the wristwatch for the people was not a technical tool where precision in hundredths of a second is important. The wristwatch, especially for women, is a prestige object whose appearance and brand are predominant compared to the accuracy.

So happened, contrary to our predictions, that the Swiss mechanical watches had not lost any market share, those quartz watches were quickly monopolized by the Japanese as cheap watches and Microma division had to be closed. Big lesson!

Palma De Maillorca. June 10, 1984. Intel sales meeting. My wife Eva wearing a Microma Lady quartz wrist watch produced by Intel

I find also a note of the sales manager Mike Markkula's speech who claimed how the solid state memories were reaching a cost per bit equal to a quarter of that of magnetic memories. Practically this cost had stood at 0.05 cents per bit!

Even Ed Gelbach added a message urging us to sell mainly EPROM Memories 2107A (rewritable ROM, invented by Intel) on which Intel was a sole supplier.

There were also many other topics like software for microprocessor, a new PLM language, development tools, etc.

In that meeting, under the beautiful sun of Palma de Mallorca, it was profiling the future and the big winning Intel.

Just over a year later, exactly from 15 to March 17, 1976, another beautiful sales meeting was held in the Swiss ski resort of Zermatt, below the spectacular Matterhorn.

On that occasion I had to have the camera handy for the many photos that I find in my archive.

Looking at the photos I understand the reason of all these photos: the camera I had brought with me wasn't there for picking up business activities but to photograph my family, then grew up with two small brats and so it was for them that I brought it.

Business wise I read a statement from the Jack Carsten's speech, the new sales manager after the departure of Mike Markkula, advising the salesmen present there to be careful on taking orders for too far ahead because we were in a "seller market" and prices were expected to rise.

I remember my two brats of almost 3 and 4 years that during the meeting they were acting like little rascals. Tom Lawrence, who was present and who could know them well, called them small "mobsters", a term that I did not know and that I searched on the vocabulary without finding it (there was no Wikipedia!). I could only understand the term while in the United States and I found the term as perfectly appropriate.

Zermatt, CH. 15-17 marzo 1976. Intel sales meeting. I was accompanied by my wife and two small "mobsters" as Tom Lawrence called them

I am sure that that wonderful resort in one of the best skiing location in Switzerland was chosen to give room to two great skiers such as Bob Noyce and Tom Lawrence ... I would only stick to the white slopes, at best!

Also on that occasion, between skiing and other amenities, sales and technical information were given in great details.

My notes show the comments by Dave Williams, marketing manager of MCS components (Microcomputer Systems), which announced a turnover of over $3 M for the first time.

A 5 volt version of the microprocessor 8080 was announced with the name of 8085, and we have been told that today microprocessor's peripheral chips would be compatible with the next 16-bit generation.

Some preliminary speech about the next 16 bits microprocessor and the N-channel technology seemed to me worth to take a note.

Several other components were announced that each of us had to start selling, is my last note. The Intel catalog was getting really big!

This is followed by some photos taken during this meeting in Zermatt and then follow other nice pictures of meetings with Intel, meetings, both business and familiar.

This meeting in Zermatt was followed by many others and always in beautiful places and with days divided between sales and technical lessons with visits to local places and nice shows ever offered by Intel.

Having represented other companies such as RCA, National Semiconductor, Teledyne Philbrick, HP and many more, I attended at their sales meetings too but I must say that only in a few cases I found the atmosphere that I perceived at Intel.

I can't describe them all here, wouldn't be enough the entire book, so I step in bringing back a few of other nice pictures of meetings, both business and familiar.

I'll limit myself to a brief description in the caption. I hope you will found enjoyable and curious this interval.

Zermatt 15-17 march 1976. Intel sales meeting. Dr. Robert Noyce, founder and Intel president... and great skier

Zermatt 15-17 march 1976. Intel sales meeting. Ed Gelbach, Intel VP and world marketing manager

Zermatt 15-17 march 1976. Intel sales meeting. Tom Lawrence, Intel European director... and great skier

Zermatt 15-17 march 1976. Intel sales meeting. Stan Mazor microprocessor training manager and co-inventor of microprocessor

Zermatt 15-17 march 1976. Intel sales meeting. Da sinistra Gert Griese, Guy Debruyne, Ms. Cooleman, Ken Boyce

Italy, Sicily, May 1975. A very young Stan Mazor with his beautiful wife Maurine in a holiday break after teaching Italian customers on the use of microprocessors.

Santa Clara 1980. At Intel with Jack Carsten reading an Italian magazine speaking about Intel microprocessors

Palma De Malliorca, June 1980. Dick Clover, magnetic memory director, showing a new magnetic chip. This technology was abandoned by Intel

Paris, Lido. June 1980. A great evening after a day work with Lou Calcagno (Intel South Europe) and Bernard Giroud (France) and his lovely French wife

Palo Alto. Nov 1977. Visiting Hal Feeney (Intel system manager) and his lovely wife after a day at the Santa Clara Intel factory

Santa Clara, June 28, 1991. At Dr. Gordon Moore's office with Ms. Jean Jones, his secretary, while entertaining my four kids Emanuele, Edoardo, Elena and Enrico, now grown up quite a bit!

Santa Clara, Nov 22, 2008. And so many years after my beautiful story with Intel, with my last son Enrico (Henry), I went back to revise the great factory dedicated to the disappeared Robert Noyce and to visit the internal museum. I could describe much of what we saw in

the Museum and with a certain nostalgia. Now Henry became an American citizen and lives with his family in Texas.

1988. Visiting Santa Clara and reviewing the great Ed Gelbach remembering the many and sometimes tumultuous, but always constructive meetings at Intel

1988 Los Altos. Visiting Stan Mazor's family home

Semiconductors: Military and Space Race

This story took place during the hottest years of the cold war, in the second half of the last century, and I would like to report in this technological history the environment in which I lived then.

These were the years ranging from 1960 to 1970 and raged a space race and a race for military supremacy between the USSR and USA. Here follow some facts that impressed me a lot.

Ruled the cold war, Vietnam war was raging in the far East and America was immersed in a frantic military and space race with the Soviet Union

This is the time when everything is intertwined with the development of our integrated circuits: these components had a key role in the race between the two super powers and a small part of the US success in that race is due to them.

I followed the events with interest and fear. Those were times when I had read that a nuclear surprise attack could be countered only if within 5 minutes there was a reaction. But in five minutes we could barely make the sign of the cross!

There were talks about intercontinental missiles, multiple-warhead missiles, nuclear submarines that each one alone could destroy half of U.S., nuclear tests were announced daily, and these tests were pouring radioactivity into the atmosphere. In short, it was a continuous psychological bombardment on how we should die soon for some frightening event. All this is described in a 1963 wonderful movie by Kubrick entitled "Dr. Strangelove".

You will ask what all this has to do with our Monsterling: well, dear reader, it has had a very important part at that time, a submerged part that is not recognized today like it should be.

And they had to do also for the conquest of space, first won by the Soviets with the dispatch in the of Sputnik, the first artificial satellite, the dog Laika, first living in orbit and Gagarin, first man to cross the door out of our atmosphere.

It was our own Monsterling and his children who then allowed the West to bridge the gap with the Soviets and then to overcome them by bringing the first man to the Moon in July 1969.

"How could the Transistor be so effective?", you will ask a bit incredulous.

Well, it was the miniaturization of electronic devices obtained by using Transistors and integrated circuits that the USA has quickly been able to design and build much lighter and more reliable space ships than those the Soviet Union still using vacuum tubes in their electronic systems.

It is so that while the gigantic Soviet missiles brought in orbit satellites with antiquated technology of enormous weight, unimaginable for the much smaller western missiles, the U.S. was able to send in orbit efficient satellites of modest weight and full of semiconductors consuming less electric power.

In that period, in fact, it has been thrown the industrial foundations for those innovations that will lead us to the present day, and that in the meantime will allow the Americans to win the race between the two blocks.

The West was quick in the industrial development of increasingly smaller and powerful instruments based on these new technologies and to allow the Americans to obtain the great results that we all know today.

I still have many notes, some particularly curious other funny. What follows in the next chapter is one of them that really has not mach to do with our technology but it might interest the reader anyhow.

The Atomic War avoided: NADGE

Here's an interesting fact that I want to mention and that I found by reading my old notes while preparing this book and that gives an idea of the atmosphere in which we were living the cold war at that time.

I had read on a technical magazine that in the year 1960 it happened a fact that could have had disastrous consequences for our entire planet.

Being myself a writer I was looking for any news that might interest my readers and the news that I read and that I will report here left me and the readers with the feeling of being sitting on a volcano ready to explode. A bit like in the Kubrik's film I mentioned before.

Feel what did happen: the Americans were obsessed by the urgent need to reduce the reaction time to a possible Soviet surprise missile attack ... those five minutes within which to react to have an effective deterrence.

First they should then be able to perceive the arrival of these missiles as soon as possible after their launch.

Did not exist networks of spy satellites dedicated to this purpose, as today, and the only way to catch the launch of a missile over great distances were the ground based radar systems with the maximum available power and located as close as possible to the enemy territory.

For this reason it was installed a powerful radar system, called NADGE (Nato Air Defence Ground Environment), close to the North Pole from where it could detect any object launched by the Soviet Union territory after a few minutes from the launch and so capable to warn the US strategic commands in due time.

The whole system was governed by a complex computer network that based on a well-thought-out program would analyze all signals received by the radar and capable to report the attack. The computer program was designed so that it could distinguish an actual massive missiles attack from flocks of birds, small planes, etc.

Here's what happened: during the first month of activity, the computers had given the red alert for a gigantic missile attack by the Soviets. The program provided unambiguous data by analyzing the strong signals reflected on those that appeared to be metal missiles just taken off from the Soviet territory.

In the same magazine was reported that fortunately a technician at the Arctic base, suspicious for the strangeness that just during the first month of operation of the radar there was already an attack, with a providential manual verification, he could verify that no attack was underway and that the alarm was false.

What did happen? Those radars were so powerful that when the moon rose over the horizon reflected back the signals coming from the radars, and the computer network interpreted those signals as they were coming from a myriad of intercontinental missiles in flight to the USA!

What a risk, boys! The computer programs had not yet been taught to distinguish the Moon from the missiles and that clever technician saved all us!

I will report in the next chapter another curious news that I read during those years referring to a military exercise where aircraft were supposed to fight in an radioactive environment like the one after an atomic explosion. In that environment our semiconductor devices have shown all their weakness.

The Vacuum Tubes win

While in the United States in the late 1960s virtually all avionics was based on semiconductor circuits with the benefits offered by their small size and high reliability, in the Soviet Union was still widespread the use of old technologies based on valves, also on board of planes.

Well, what had happened and what brought me one of the many magazines that I was following and of which I took note for an article?

During a training exercise in which were tested aircraft in a combat situation inside a radioactive environment, probably as a result of a nuclear test, the American fighter planes had shown to be extremely vulnerable and severely undercut by the Soviets planes.

A Pentagon investigation found that due to the fact that the Soviet avionics was based on old radio tubes, insensitive to radiation, avionics with valves was much more reliable in that situation and much less sensitive to radiation than the American avionics based on semiconductor electronics.

I don't know much more, but I remember that in the late sixties in USA were advertised "radiation hardened integrated circuits", namely radiation resistant and I remember that even dedicated factory to produce them was founded in the Silicon Valley. Our Monsterling had been hooded in a suit anti radiation!

Until 1970, there were continuous nuclear tests

Even in the space race the semiconductors played their good role, filling the space shuttles and even descending on the moon.

Not least, in some Apollo lunar expedition, the Intel microprocessor i4004 was used onboard Lunar module, and its seats now on the Moon ground.

In the mood of curiosity, here is another gem of which I has been a witness: the mystery of the "soft error".

The Mystery of Soft Errors

This is another mystery about semiconductors and that was resolved around 1971, just when Intel started supplying the largest memory for the time: the i1103, a random access memory containing 1024 bits.

What kind of memory was this i1103? When Intel succeeded to produce it in 1971, this was the first semiconductor memory capable of competing with the magnetic memories used in computers and had a cost per bit compatible with the memory market.

Intel made a great fanfare around it. I remember an advertising page in which Intel showed magnetic memories in mourning ... and really the computer industry started to replace the old magnetic core technology with this brand new and reliable integrated circuits.

This advertisement appeared in 1971 in conjunction with the announcement of the first 1024 bit RAM i1103

I proudly presented to important Italian customers this memory and especially to Honeywell and Olivetti which together accounted for then well over half of the entire Italian market.

But before talking about this soft error I have to make a hint to the physical operation of this memory, very innovative operation for those times, but that would have reserved soon some surprise.

In practice, the one and zero of each bit were represented by the presence or absence of a microscopic electrostatic charge which was retained by a very small three-Transistor circuit.

This charge, however, faded in a time of a few milliseconds for which problem inside the chip existed another circuit whose function was to refresh every 2 milliseconds the charge and in a continuous manner: for this reason these memories were called dynamics.

All of this may seem strange today, when these dynamic memories have been replaced by much more reliable static memories, more efficient and much safer.

Let's come to the amazing and mysterious problem that Intel soon had to remedy, a mystery which I had to deal with personally and with a mixed sense of wonder and fear.

Honeywell was very interested in replacing the core memories inside its mainframes, big computers at that time, with these reliable and robust semiconductor memories and I had supplied them a certain quantity to allow them to perform all needed tests and to develop the first computer prototypes with them.

Needless to say that I personally and all Intel were anxious to carry out these tests positively and therefore we put special care for these first supplies; all the parts to be delivered were tested and retested to ensure that everything went well and I, in contact with the customer, was following the progress by speaking and visiting the various technicians at Pregnana, near Milan.

At first everything was perfect: each chip was received by the customer and individually checked to verify that the features correspond to the parameters provided by Intel.

Not only it was verified that the products conformed with the promises, but always, at the customer's site, the test results were broadly better than those promised by the manufacturer, especially as regard speed and power dissipation.

My written reports, which regularly I was sending to Intel engineers by telex, confirmed this continuous perfect proceed and I thought the matter to be going really well: surely the realization of prototypes would work fine and Honeywell would pass then to mass produce their computers and to order us substantial quantities of that memory.

While everything seemed to flow smoothly and I was already smelling the next big orders that for those times seemed very large, I started to receive strange signs of perplexity from Honeywell technicians.

I didn't get it right; from an open and almost daily communication I perceived a continuous increase of skepticism without a clear explanation.

The thing begins to get clear when they called me in the factory to give me back few i1103 chips that they claimed to contain some flaws.

Of course I did not consider the thing serious: it is normal that on the thousands of components delivered there could be someone who had shown some weakness or who had failed for some reason.

I sent these chips to the factory, so that Intel could analyze them and to find a remedy and I was waiting, with much anxiety, the diagnosis that then I would have to transfer to my client.

You can imagine my surprise when I received the diagnosis that was saying: "memories received work perfectly and according to our specifications, it is the customer who is mistaken, however we replace them just in case".

Back then to the customer and with all the precautions of the case I did observe that maybe there could be some flaw in their diagnostics and that however we replaced free of charge all the returned parts.

I thought that I passed that dangerous impasse, but it wasn't so. After several days they recalled me at the factory for the same problem and I had to return to Intel another set of faulty chips as per client's claim. Needless to say, I made the same circle: sent them to Intel, Intel analysis confirm that the parts are OK, replacement, my return to the customer.

If one that story hadn't lived like me I think he would have trouble to believe it: we went on like that for a few months and I was in the middle; I didn't know what to believe, whether the fools of the village were at Intel or at the customer.

Fortunately, as I learned later, the same thing was not happening only to me, so at Intel technicians began by taking seriously the phenomenon and not just by considering my client and myself as a bit crazy and they started trying to figure out if there was some truth in all that strange story of evanescent defects.

I was informed that also at Honeywell they began to investigate those unexpected phenomena that were generating those strange errors and that had no identifiable faulty chips.

I remember that the explanation came from Intel as a lightning in a serene skies after about six months from the beginning of that hustle.

I was told that the culprits were cosmic rays! Yes, you read that right, cosmic rays, those rays coming from the stars crossing the atmosphere and someone hitting that microscopic electrostatic charge contained in the chip. By doing so those rays were discharging the static charge and changing a one into a zero, triggering an error signal on the computer.

Remained to explain why only Honeywell had this damn problem and no other Italian customer. And here I was given an explanation which then also Honeywell confirmed to me.

Honeywell was the first customer to substitute big core memories with those chips (i1103), whereby the probability on a large plate that contains many chips to intercept a cursed cosmic ray to create that mess was high. More, Honeywell was testing its computers for days and the probability to catch a cosmic ray was 100%, and we have seen that fact.

That's why testing the individual chips and for a short time it was impossible to fish that cosmic ray. These stupid rays were simply discharging that electrical charge and not causing any other damage, so their wrong behavior could not be detected in retrospect: for this reason it was called "soft error", soft or sweet, I would've called it damn error!!

This problem was quickly remedied by shielding the chips ... a bit like it was done with semiconductors affected by radiation from atomic explosions, as seen in a previous chapter.

At the conclusion of this story which really made me fear the worst for this technology, still in its infancy, I must add that broader and general research carried out in the presence of other soft errors found that in general the chip is sensitive to neutrons also coming from earthly sources and the chips must be shielded.

The Transistor's Siblings

Once disclosed the solid state physics, many scientists have indulged by inventing dozens of other strange semiconductor components.

Many have aborted soon after been invented or for their impossible applicability or for their excessive production costs.

In the articles and in the news that I was publishing in those days I too had fun in describing them for the joy of the readers and having kept copy I can reread them today.

I quote below some name from my published news and I may go on with dozens of strange semiconductor devices, but I limit myself to just a few.

Field Effect Unipolar Transistor (FET), Metal Oxide Semiconductor (MOS), Complementary Metal Oxide Semiconductor (CMOS), Thyristor, Dinaquad, Tunnel diode, Light Emitting Diode (LED), the Franch Tecnetron, Silicon Controlled Rectifier (SCR), the German Transitron, the Spacistor, the Nuvistor, Spacesaver, Semitron, Thyratron.

Some have remained and developed and became the basis of modern electronics and an important part of the next chapters, where I'll provide extensive information. Two will become the bricks of modern electronics: FET, acronyms of Field Effect Transistor and MOS-FET acronyms for Metal Oxide Semiconductor-Field Effect Transistor.

To satisfy the historical curiosity of my readers I scanned some pages of the old documents that are worth seeing now.

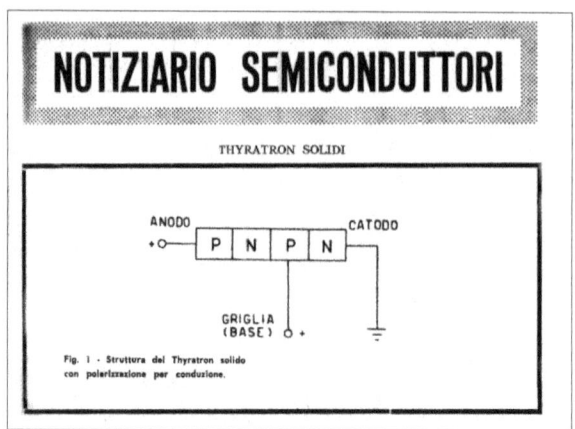

Dec 1962 – My "Semiconductor News" about the Thyratron, a multi-junction device aimed to be used in power switching controllers.

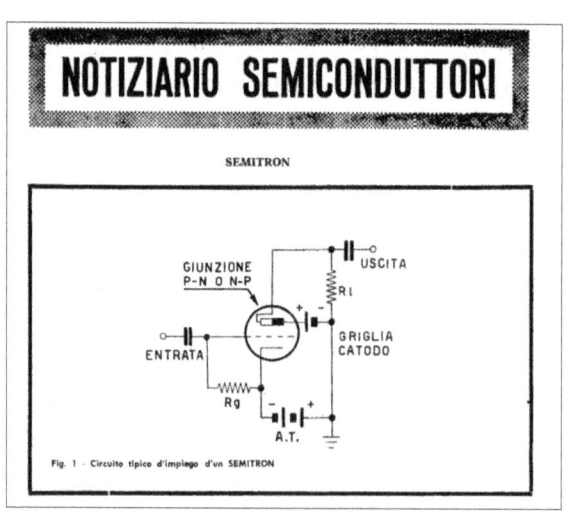

Oct 1962 – My "Semiconductor News" about Semitron. Half vacuum tube and half semiconductor. An hybrid combination where the designer have been looking for to obtain the best from a semiconductor junctions and the best from the vacuum tubes

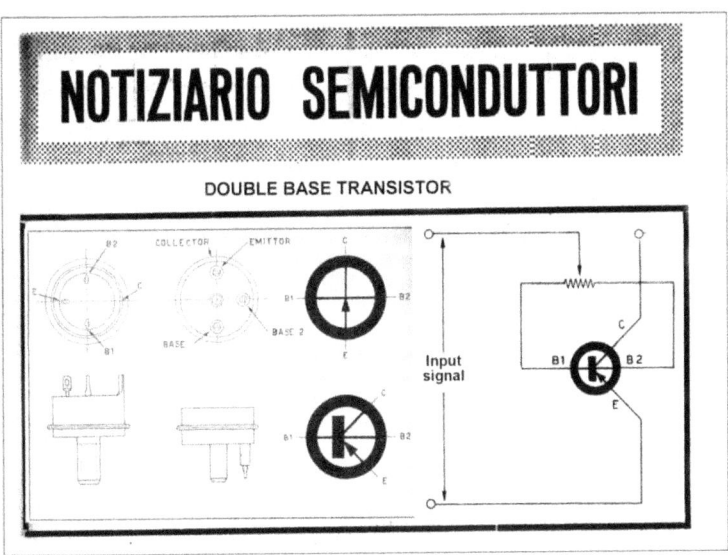

Sept 1963 – My "Semiconductor News"on Honeywell tetrode power Transistors. The two bases were designed for the output stage of fully transistorized Hi Fi amplifiers. By modulating the second base it was possible to linearize the output and to come close to the much better distortion of vacuum tubes I bought and tested one of these Transistor and I should say that its performance in an amplifier were far better than those of Philips Transistors at similar power.

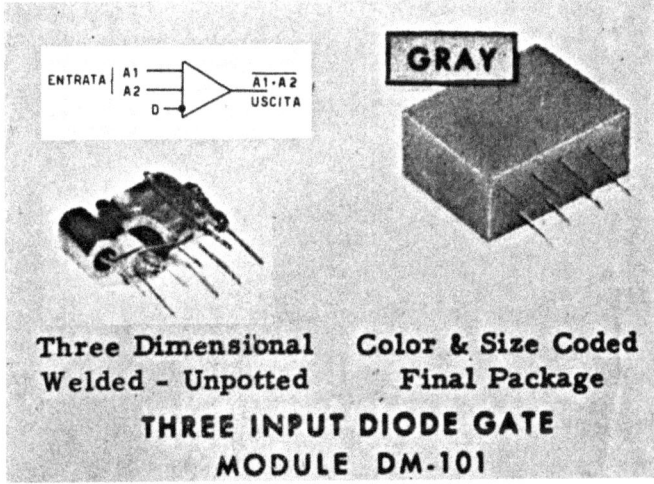

Late 1950s. A solid state logic circuit made by assembling discrete components and encapsulated in a plastic package (integrated circuits didn't exist yet)

Solitron's new high power switches are NPN Silicon Devices. These devices are designed to replace many parallel low current devices used to obtain fast switching at high currents. They are ideal for use in high frequency pulse generators, high voltage switching circuits, solenoid control circuits, and other circuits requiring high speed switching devices.

Early 1960. A high power transistor for the time

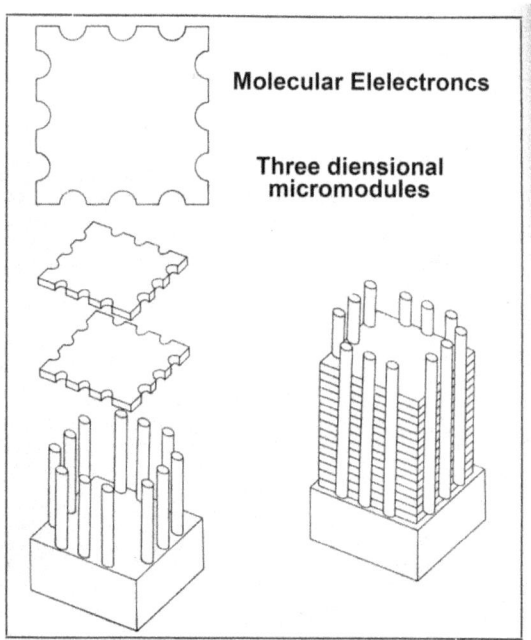

In the mid-1960s I received a strange documentation describing a technique, never heard before, called "Molecular Electronics". Practically consisted in the vertical assembly of small modules containing an integrated circuit. The aim was the realization of extremely compact and reliable electronic systems to be used in the military and space instrumentation.

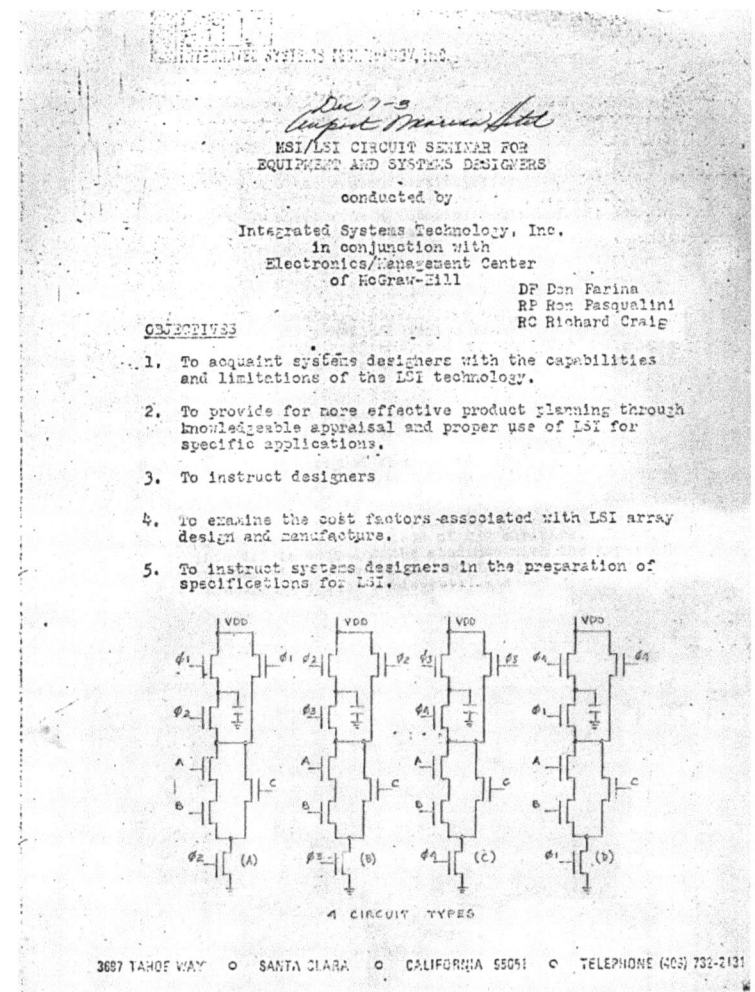

Santa Clara, late sixties - This 12-page historical document came to me from Fairchild. This copy is originally signed and refers to a Seminar addressed to users of MSI and LSI integrated circuits. This conference was held by Dan Farina, Ron Pasqualini and Richard Craig in 3697 Tahoe way, Santa Clara, and was sponsored by Integrated Systems Technology inc. and the Electronics Management Center of McGraw Hill.

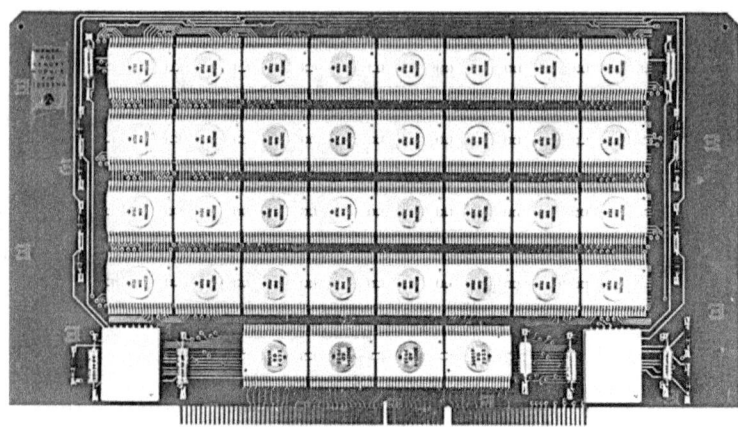

1969. Rockwell Memory Module: 2048x8 bits, cycle time 1 microsecond, read access time 500 ns, size 6.5"x12.0", price on request!

1969 - Anaheim, Ca. An I. C. assembly line at North American Rockwell, Microelectronic Division

1969 Anaheim, Ca. North American Rockwell, Microelectronic Division. A lady, technical operator, in the testing department of integrated circuits. Fashion has certainly changed since then, as well as the electronic components!

1960 – 1970. Some of the envelopes that I have retained coming from American companies and full of technical documentation

I think interesting to bring back here a note that I read on the Rockewell documentation that I received then. The note states:

" The Minutman I, II and III programs, perhaps more than anything else, led the way to widespread the use of advanced microelectronics for industrial and consumer products".

A frank and important confirmation how the military industry has been key for the development of solid state electronics in the United States.

In this chapter I have been describing for young readers and mature readers the beginning of the developments of modern electronic also providing some original document from my archives.

This is a proof of how many brothers, nephews and cousins have immediately been of our young Mosterling, relatives who have found some immediately a job, others who are now retired and many who have not been successful at all.

Among these who have succeeded indeed, a sensational success, are the FET and the MOS-FET Transistors which we will analyze in detail the operation in the next chapter.

Field Effect Transistor (FET)

This is another important invention of William Shockley published in 1952 in the Proceeding of the Institute of Radio Engineering (today IEEE).

The article is entitled the "Unipolar Field Effect Transistor", as he called this component, and it is the ancestor of future Transistors like MOS and CMOS, the main components of modern integrated circuits like microprocessors and memories.

Here is the FET, a new Monsterling, the younger brother of the first metallic Monsterling

A Field Effect Transistor (FET) operates on a principle different from the one we learned in the chapter dedicated to the junction bipolar Transistor but, incredibly, its theory can be traced back to the 1932, when someone thought and patented his working principle.

Its physical principle is quite simple to understand, but only the discovery of the phenomena involved in semiconductors have enabled Shockley its practical implementation.

What had been thought in 1932? To apply a transversal electric field to a piece of conductor throughout which was running an electric current.

Modulating this field was thought to be possible to modulate the current just as in a vacuum tube triode where the flow of electrons between cathode and anode was modulated by an interposed grid.

The idea was very logical but still they tried to modulate the current they could not achieve the desired effect.

At that time various materials were tested but however strong was the transversal field the electrical current running throughout any material tested did not change at all.

Only with the in-depth knowledge of the behavior of semiconductor materials, such as germanium and Silicon, it has been possible to obtain an useful transversal modulation and thus a Field Effect Transistor has been created.

The failure of the 1932 experiments are due to the fact that a conductor has a huge amount of mobile electrons, while a semiconductor much less, billions of times less, so in a semiconductor material is much easier to modulate its few mobile electrons, provided that some superficial effects is eliminated, as Shockley understood and identified with mathematical precision.

This model of Transistor, i.e. components that are based on the field effect principle, have become the most popular in the industry.

As we will explain in this chapter prevents this Transistor is very suitable to be used as a quick switch, i.e. work between the two states of zero and one.

As we know in all the computers and in all the digital tools the basic component is a switch that can take the two states of ones and zeros.

With field-effect devices we leave the world of amplification and we enter the digital world, much bigger.

Today it is normal to build chips that contain billions of these switches and no one can see the final limit yet.

We will now study how this FET functions as a switch. To do that we must return to what we studied in the chapter dedicate to the diode.

As we recall, the diode is formed by a P-N junction or an N-P junction and its peculiarity is that it lets the current to flow in one direction and not in the opposite direction.

Exactly we say that the diode is forward biased when it's conducting and we say forward biased when it's not conducting.

Let's create a P-N junction with a shape as shown in the next figure where, in particular, the area N has two electrical connections.

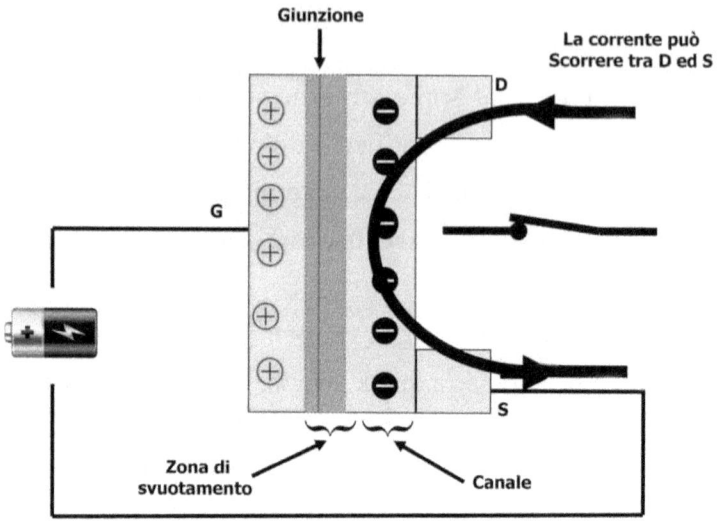

If the FET is not polarized, the channel is full of electrons and therefore there is conduction between D and S

These three electrical connections have been named:

G = Gate

S = Source

D = Drain

So the N zone, the one that has negative charges, is now connected to the two terminals D and S.

Now that we have this strange diode, we can ask: what happens between D and S".

Follow me carefully: the N zone, as we know well, is full of mobile electrons, so between D and S there is conductivity (this zone is called also "channel"), i.e., if we apply a battery between D and S an electric current will flow.

This means that it is as if between D and S there is a closed switch as shown in the figure. So when the FET is not biased it can represent what in the digital world is a "one".

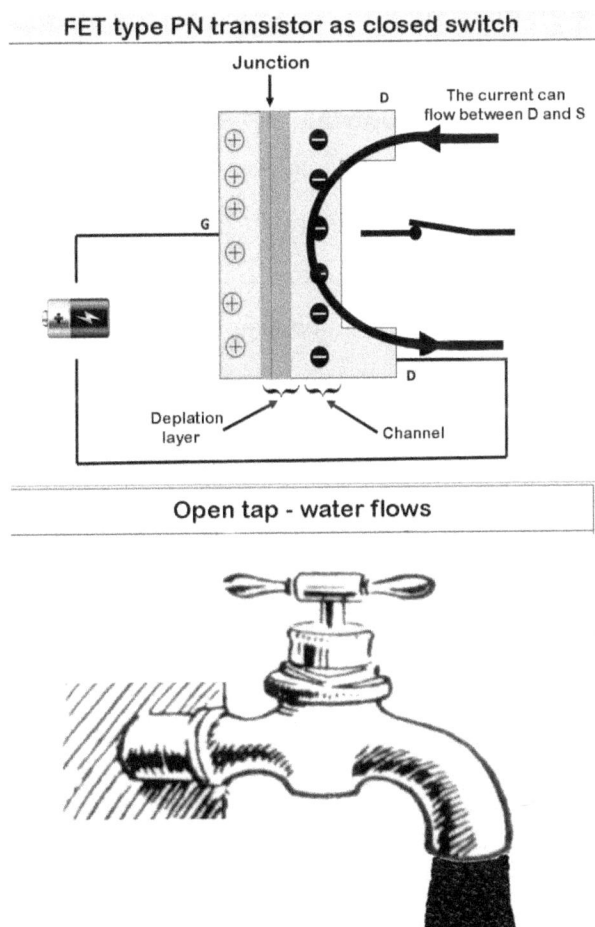

If the FET is not polarized the channel behaves like an open tap that lets water pass

The explanation doesn't stop here, and one may already understand that if we connect a battery to our semiconductor structure something different is going to happen.

And in fact something else happens. Let's keep in mind the depletion layer, that area without mobile electric charges around the junction: we know that this area is not fixed or, better, we can widen or shrink it at will by applying an electric field that introduces charges into our strange diode.

If, therefore, the zones P and N have been suitably prepared, as Shockley described in his papers, by applying a forward biasing battery then we can manage to enlarge the depletion layer to cover the entire N region. Thus the N region, depleted by mobile electrical charges, becomes an insulator.

Between terminals D and S we have therefore an area that being devoid of electric charges (electrons in this case) works like an insulation and between D and S cannot flow any current, just as if we had opened a switch or closed the faucet and stopping the water.

In the digital world you would say that this Transistor is in the "zero" state.

Let's carry on and now we connect a battery to our FET between G and S, the diode is so in reverse mode, i.e. positive on zone N (terminal S) and negative pole on zone P (terminal G).

From what we learned about the diode, we know that by biasing it in reverse mode its depletion layer widens.
The higher the voltage of the battery, the wider the depletion layer.

There is a voltage at which the depletion layer will widen so much to deplete completely the N zone from its electrons.

But we also know that an area without free electrical charges is practically an insulator and therefore it will not flow any electrical current between terminals D and S.

We can thus affirm that in these conditions the FET acts as an open switch, ie, in digital terms as a "Zero".

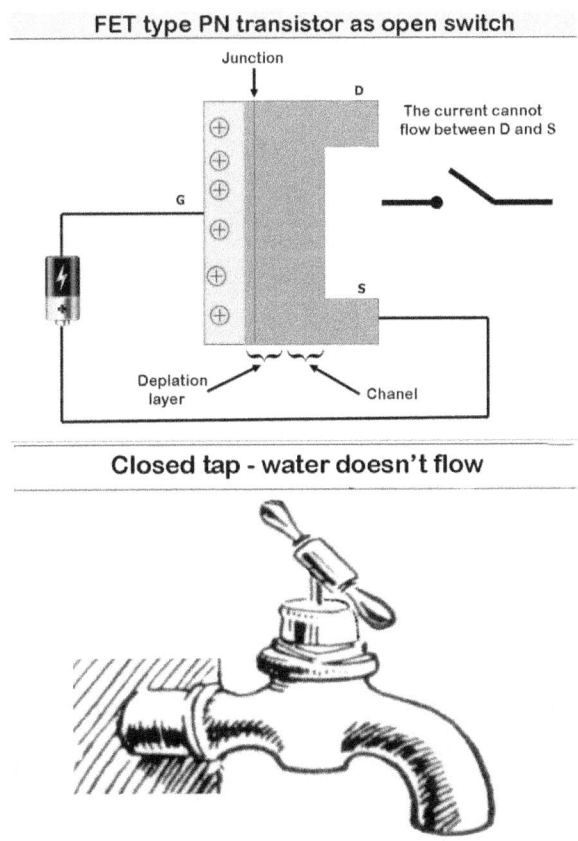

FET type PN transistor as open switch

Closed tap - water doesn't flow

If the FET is polarized the channel acts like a tap closed and do not let water to flow

But we also know that an area without free electrical charges is practically an insulator and therefore it will not flow any electrical current between terminals D and S.

We can thus affirm that in these conditions the FET acts as an open switch, ie, in digital terms as a "Zero".

We have thus discovered how to exploit this FET as a switch in any digital systems and hence its importance.

In a modern personal computer, when we type on the keyboard, inside the computer these switches open and close in millionths every second.

To complete this topic, let's see the electrical symbols used to indicate these components.

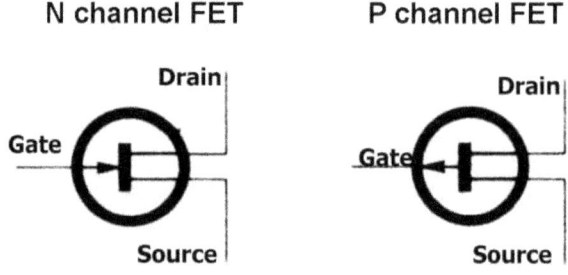

FET electric symbols

The FET, and in general the field-effect devices that we will see in the next chapter, may seem a modest variation of the junction Transistor, nevertheless it has become the most widespread electronic component on the market for its simplicity of construction and for its excellent electrical characteristics that make it a quick switch and a very low power hungry component.

In addition this technology is today present in all the digital cameras on the market where the sensor is measured in megapixels and each pixel is nothing but a field effect Transistor (MOS-FET to be precise). A 20 megapixel camera, now standard, so contains 20 millions of these components.

Transistor MOS -FET

At the beginning of the book we have often cited the germanium as an element to get the Transistor effect.

Now we will deal with a field effect Transistor made only of silicon because it must be known that practically in the first half of the sixties the germanium remained only for the realization of very special individual Transistors.

With the advent of the integrated, silicon enforced its superior characteristics over germanium in terms of productivity, consumption, thermal stability and, above all, integrability.

Today it would not be even thinkable to produce a germanium microprocessor.

More, can you imagine to call the today "Silicon Valley" "Germanium Valley"?

MOS-FET Transistor funny character, young Monsterling's brother

In the 1960s, with the advent of the integrated circuits, showed up the heating problem. By increasing the number of components crammed in the same chip also the power consumption was increasing and therefore the chip could burn out.

It was important to manufacture Transistors less power hungry than the junction ones and, as seen in the previous chapter, at RCA the first Large Integrate Circuit (LSI) very power thrifty was invented, precisely the CMOS (Complementary Metal Oxide Silicon) that took advantage of the FET-MOS Transistors.

With this New Transistor became possible to integrate very large number of components in the same chip, impossible before.

Still today this is the type of Transistor that is used in most microchips. Over time it has been refined and many variants have been introduced.

The name comes from the fact that a thin metallic layer was inserted over the channel isolated from it by a layer of silicon oxide, hence the name "Metal oxide Silicon".

What follows are the electrical symbols of the MOS-FET P-channel and the MOS-FET N-channel.

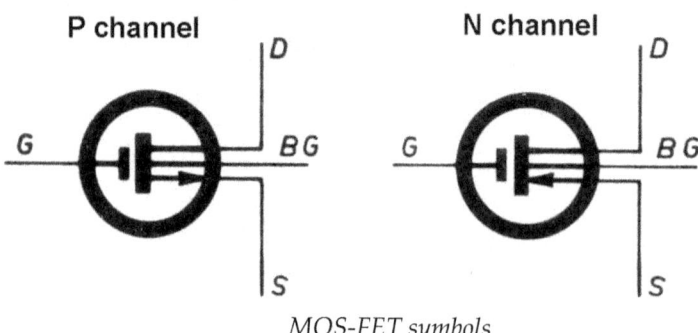

MOS-FET symbols

Its operation is conceptually identical to that of the FET junction already seen. It is based on the ability to open and close the channel underlying the metal layer as a very fast switch and with extremely low power consumption, thousands times less than a junction Transistor.

In addition we should mention, that by reducing more and more the channel size, the industries were able both to increase the speed and to pack billions of them in a single chip.

We have thus maximized our capability to take advantage of silicon, one of the most abundant elements in nature, while the rare and costly germanium have been left for very special applications.

It is a common mistake to think that silicon is so common that its cost to manufacture electronic components should be zero

Unfortunately it is not so because the microchips are manufactured using very pure monocrystal of silicon.

The process to create these monocrystal suitable for use in electronics is extremely expensive and an ingot, like those used at Intel, can cost even hundreds of thousands of dollars.

A clarification: we here have been speaking of MOS Transistor, while in the specs of your camera you find the term CMOS, which stands for Complementary Metal Oxide Silicon ", the acronym seen that refers to a construction method of the circuits that connect the various Transistor.

Finally for technically expert, I should add the actual reason why this MOS-FET exhibits its special electrical characteristics.

We should refer to the equivalent input circuit shown in the diagram below.

In this chapter we explained that the MOS-FET is made by inserting on top of the channel a metal layer isolated from the channel by a layer of silicon dioxide, i.e. glass.

This set is nothing else than a capacitor inserted at the input of the MOS-FET.

This capacitor is practically an insulator for which the MOS-FET differs from any other component, both solid and vacuum, due to its particular characteristics that allow circuit simplifications not achievable with other devices as there is no input bias and it is thermally extremely stable

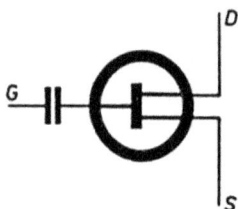

In fact, the channel zone is immediately under an insulating layer of silicon dioxide and its conductivity depends on the voltage applied to the electrode G.

When an electric current flows between S and D through the channel zone, any voltage applied between G and S generates in the channel a field effect that changes the state of the mobile charges and then modulates the current flowing between S and D

In addition, in production, it is possible to modify the structure of the electrodes, and to obtain two fundamental types of MOS-FET, respectively, called "enhancement" and "depletion", which differ on how they behave electrically.

The "depletion" type can function with grid bias zero, that is, as if you had a junction Transistor whose best working point is at zero current in its base, practically with no power consumption.

Contention Texas Instruments-Fairchild

As we know, the integrated circuit was born in the late fifties, precisely in 1958 by the work of Jack Kilby of Texas Instruments, but much of the credit must also be attributed to Robert Noyce of Fairchild Semiconductor, the latter is the founder of Intel in 1968 along with Gordon Moore.

What happened to this invention and why the Nobel Prize was awarded to only Kilby in the year 2000? And while the invention is also credited to Noyce?

This chapter is dedicated to what has happened on this point and has caused tensions and lawsuits for many years between Texas Instruments and Fairchild.

To understand this mess we must briefly go back to when the Transistor was discovered in 1947 and what followed.

The Nobel Prize for the Transistor was awarded in 1956 to equal merit among the three who worked there while, in our case, this did not happen: we will see why.

Let's start by remembering that it was Shockley that started what will become the Silicon Valley, by founding in 1956 his company, Shockley Semiconductor laboratory, in Palo Alto and hiring eight scientists among whom Bob Noyce.

These eight and Shockley were then the only one in the world with the know-how on the production of solid state silicon components and the year after the eight left and founded the Fairchild Semiconductors.

At Fairchild, they were so far ahead in the use of silicon and already produced transistors with planar-type techniques, that is, the technique that will be used to produce integrated circuits till now.

In 1958 at Texas Instruments, a young technician named Kilby, using a laboratory left free for the holidays of the other scientists thought to connect some components created inside a germanium crystal.

After connecting them with gold wires, the circuit, an oscillator, worked perfectly as if it had connected the separated components.

Returned from vacation his chief thought that that strange system was worth to apply a patent for, which they did.

At the same time at Fairchild Noyce was applying the planar process, invented by his colleague Dr. Hoenri for the production of reliable silicon Transistors, to connect several components diffused in a slice of silicon monocrystal.

The thing succeeded perfectly and thus was realized the first real integrated circuit as we know them and a technique used even today.

The one at Texas Instruments wasn't really an IC as we mean today but at TI they have been good enough to patent it slightly before Fairchild and that's what count for the patent office.

From this near simultaneity was born a lawsuit over ownership of the patent between the two companies that was settled after ten years with the exchange of patents.

It must be said that both Kilby and Noyce had distinct merits and they two definitely deserved to be regarded as the creators of the microchip.

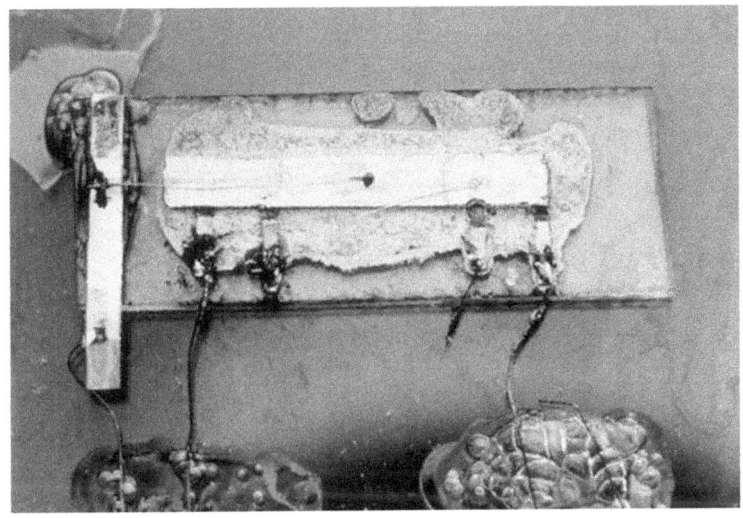

August 1958 - Picture of the first working integrated circuit developed by Jack Kilby of Texas Instruments in a germanium crystals which patent was deposited in the same year.

The same Kilby, who was awarded the Nobel Prize in the year 2000, recalled on that occasion the merits of Noyce and stated that he would have liked to share the prize with Noyce that unfortunately died prematurely in 1990.

The Nobel Prize was awarded only to Kilby and not to Noyce because the regulation of the Nobel prizes require that only living persons can be awarded and Noyce died prematurely in the year 1990.

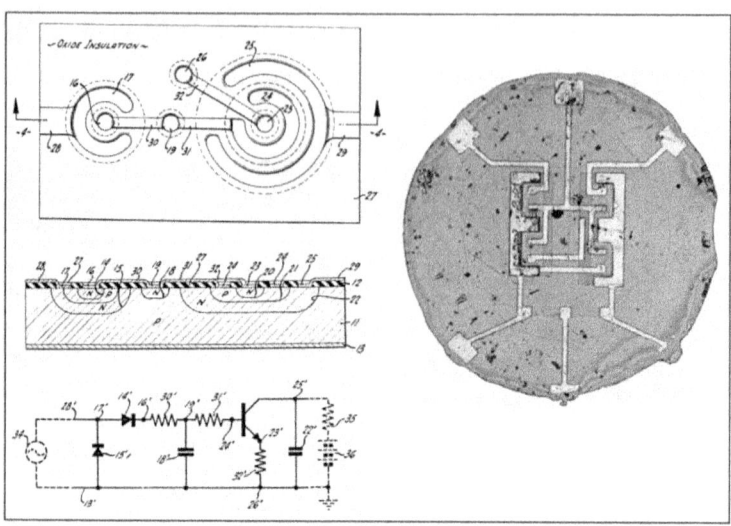

At the left three images of the patent filed by Robert Noyce in the year 1958 and approved in 1961. On the left a firs IC produced at Fairchild. This patent was deposited shortly after the one deposited by Kilby and a legal contention started between the two companies. At the right the first planar integrated circuits sold on the market by Fairchild

How the Integrated Circuits are Made

All those devices derived from the Transistor, in the last century gave rise to an incredible development of industries able to produce unimaginable products only a few years before and of which we do not see the point of arrival

If the 1947 had not opened the door towards the today miniaturization thanks to the solid state electronics and we were still designing with the valves, all those products that we use like PCs, smartphones, GPS, etc. could not exist.

Who could imagine that with the advent of those components would then be built immense factories capable of churning out billions and always at prices downhill?

All today's industrial giants like Apple Computer, Microsoft, Google, Facebook owe to those core industries, such as Intel, their existence.

This chapter is dedicated to the explanation, in a clear way for all, of how the microchip factories realize their products.

We will explain in detail, but without using specialized terminology, this wonderful world that lies between the artistic and the highly technological.

The technology illustrated here refers to the one created at Fairchild by Dr. Noyce and that is today used by all the industries in the world, whether American, Korean, Japanese, European, Chinese, etc. etc.

We already know that a microchip or chip or integrate circuit, is a set of components that belong to a single silicon crystal.

The following figure illustrates this fact by putting together a number of MOS-FET Transistors with their hat and using the funny character we have introduced in the previous chapter.

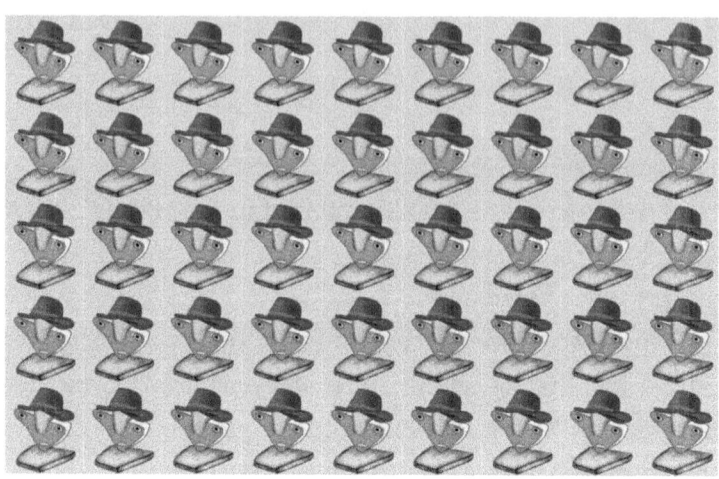

A modern chip is formed by a large number of MOS
Transistors connected in the same silicon crystal

The Noyce's production process starts from a silicon slice of very pure crystal where an insulating layer is created on its surface by oxidizing it in an oven at high temperature.

The new surface is nothing else than glass, a perfect insulator, in witch little windows should be chemically opened as we will see in the next step.

In fact Noyce thought to open in the glass some microscopic windows not acting mechanically, what would be impossible, but using special corrosives able to dissolve the glass by sprinkling it on the surface of the glass.

To open the windows only on the desired points over the glass a corrosion resistant mask is deposited with holes in the points to corrode.

In those windows the dopants are spread to form the P and N zones.

In the end vaporized aluminum is spread that settling at predermined points form in one shot all the contacts.

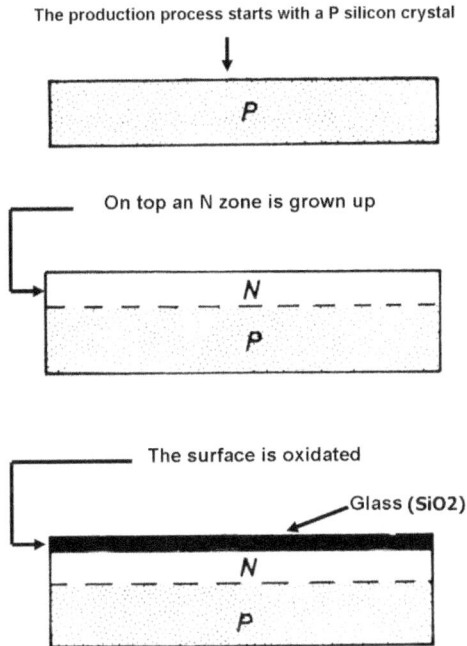

The production process starts with a P silicon crystal

P

On top an N zone is grown up

N

P

The surface is oxidated

Glass (SiO2)

N

P

First phase: oxidation of silicon to create the layer of insulating glass

Summarizing: once the surface is oxidized, it is coated with a layer that is not attachable by a corrosive acid for glass.

You remove this layer only at the points where you want to corrode the glass to prepare the next operation.

You cover the surface with acid that corrodes only the points left free and this makes visible the underlying components as seen in the following drawing.

It is here that comes the Noyce's brilliant idea: instead of inserting in each window a wire and bond it by a difficult and expensive operation, if you spread a layer of aluminum on the surface you will get that aluminum will penetrate each window and will bond the various components forming a complete electronic circuit.

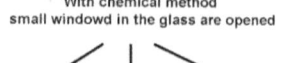

With chemical method
small windowd in the glass are opened

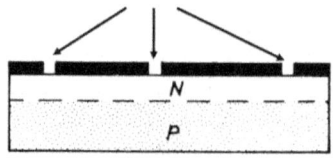

Isolation zones P are created

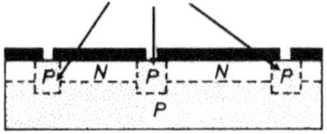

Electronic components are created

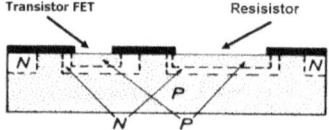

Second phase: preparation of Windows in the points to be connected

Circuit components are ready

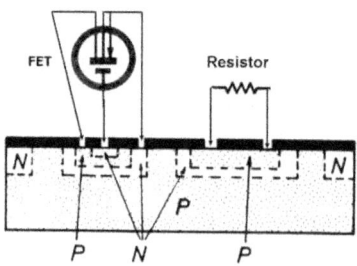

... and deposited aluminum sheet links them

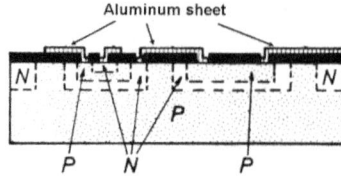

Third phase: an aluminum layer connects the various components

This method is much more sophisticated than the one proposed by Texas Instruments and especially perfect for being industrialized.

In a word, this process allows mass production of integrated circuits.

All described here is realized with photolithographic systems borrowed from the world of photography.

Extremely complex machines have been used that have reached incredible levels of precision over time.

A modern integrated circuit can contain hundreds of millions and today even billions of those small windows and as many contacts to realize, for example, a microprocessor. The result may seem miraculous if one thinks that even one mistake, only one wrong connection, would compromise the entire chip.

We should remind here what mentioned often in this book, that the exponential Moore's law published in 1975 predicts that this complexity doubles about every two years and this law has proven to be right so far.

The result is the progress we see and that we will see for many years to come ... isn't it wonderful?

Another problem the engineers faced is the heating of the chips. Increasing their complexity and their speed the heat generated inside the chip can destroy it.

A solution found to increase the performance of a microprocessor without increasing its frequency is to make more microprocessor on the same chip.

The modern microprocessor contain 2, 4, 8 and also over microprocessors in the same chip.

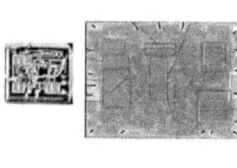
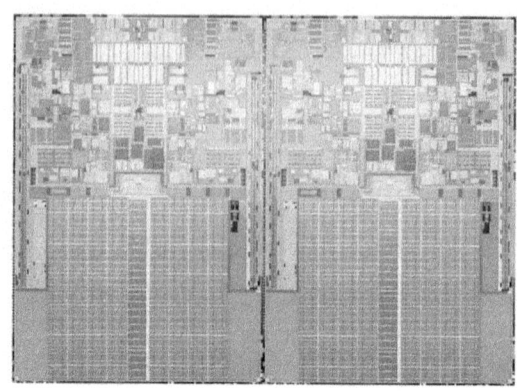

From chips of two or three mm per side and with a few Transistors the modern chips have several centimeters on each side and containing billions of Transistors

Then there is the powerful and continuous demand by users to get ever more efficient systems and simultaneously decreasing prices.

This demand is pushing the chip industry to invent new structures, new materials and new packages in which to place the various chips that also grow in size.

To meet the need to connect the chips to the outside world the packages have reached huge sizes and with a pin numbering close to the thousand

The Wonders of Production

This really is a beautiful chapter! and I am not referring so much to what you will read, but just to the intriguing and complex beauty of the factories I visited as a student.

Intel did not exist yet and I hadn't been in America but I was able to visit SGS in Italy, Philips in the Netherlands, Newmarket Transistors in England and Ferranti in Scotland and I remained impressed with what I saw.

They were my first visits to industries therefore logical that I received strong impressions, however, it was then that I understood how big a factory had to be to produce such small objects.

What I saw then has no comparison with what exists today and that I find when I visit a modern semiconductor factory like Intel, but as always, first impressions are the lasting ones.

In the previous chapter we saw how the chip and how Noyce's process, with its various steps, allows us to industrialize the mass production of the chips.

These production lines are so complex as to appear to belong to a science fiction movie. These production lines have achieved stratospheric costs of billions of dollars, and only a few factories in the world can afford such large investments.

This explains the need to find very large markets and why nowadays the companies that survive are flooding the market with hundreds of millions of pieces, see personal computers, mobile phones, smartphones, tablets and many others.

Going forward for a few more years probably some manufacturer of microchips will try to invent some product to sell it to all the 7 or 8 billion living creatures and thus to justify its future investments!

Simplifying a lot, we can compare the production of an integrated circuit to something in between a photographic technology and the preparation of a sandwich.

Yes, it's not a text copy error, just a sandwich!

The integrated circuit is in fact is a sandwich formed by several interconnected layers with the final result of a system that as a whole works.

And the technique to create the various layers is practically of photographic type: that is, the various layers of the sandwich are "printed" as they were created from a negative film.

Clear that this is about something of very high accuracy where details should be printed with a precision that now reach a few hundredth of a micron and everything has to be done by automated systems controlled by powerful computers and very special programs.

For pure curiosity the following figure dating back to the 1970 shows some chips containing several thousand Transistors and their actual size compared to a pencil eraser and a razor blade ... If a microprocessor of today is produced with that technology, then the chip would not even stay in the room where you sit now!

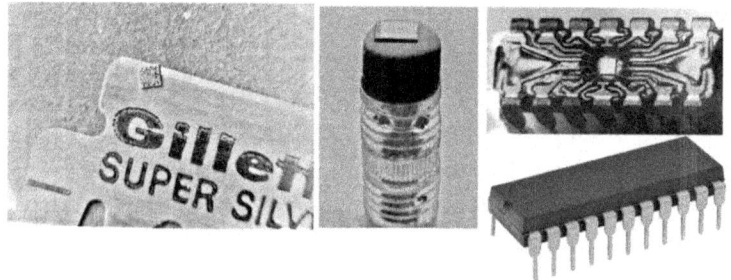

How the dimensions of the integrated circuits were shown in 1970

Today, these chips are far more complex and even containing hundreds of millions of Transistors, their physical size has not grown much, it all comes down to a few square inches.

The picture below represents a huge enlargement of a recent Intel chip on which three technicians are walking, photo taken recently at the Intel Museum in Santa Clara. The actual size of the chip does not exceed 2 inches on each side.

Some Intel technicians walking at the Intel Museum on a huge magnification of a modern chip

Over time the number of layers of the chips have increased, the processes became automated and today virtually are the computers that design modern integrated circuits based on programs that end up controlling all the production equipments.

Imagine having to design and build an integrated circuit containing a billion Transistors with the manual techniques used by Federico Faggin to design the 4004 microprocessor in the year 1971? Perhaps a century would not be enough!

And maybe you are holding right now in your hand a smartphone containing a chip with that billion Transistors as you read this book. What progress guys!

Let's see now how in the factory begins the production path starting from the ingot of silicon monocrystal.

The highly purified ingot is taken from a special machines that at high-precision slices it to like a salami.

Everything happens in an extremely controlled environment and certainly not with a circular saw like that on the figure.

*The monocrystalline silicon slices on which to produce the microchips
are sliced from a ingot as if they were slices of salami*

It should be noted that the production of these chips is made in huge series, i.e., starting from a gigantic cylindrical ingot of monocrystalline silicon and sliding it as a salami, on which slices will be literally printed hundreds of individual microchip.

As we know the production of semiconductor components takes place in large quantities and to obtain it starts from ingots with diameters up to 15 inches.

Then on each slice are "printed" hundreds chips that will be cropped in order to continue their journey in the production chain.

Each chip is tested before being sent to the final part of the production chain, as shown in the next figure.

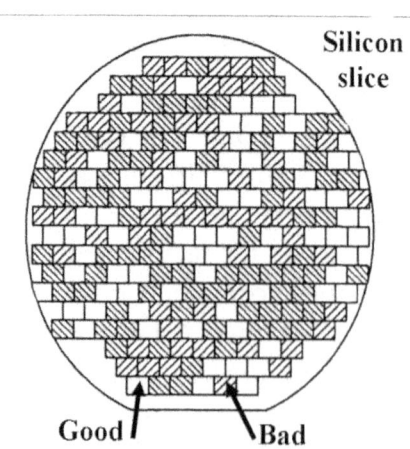

The faulty chips are marked and then discarded

At the end the good chips are placed in their respective packages, tested a second time, and sent to customers.

To complete this topic we must also know another important part of the production of our chips that we flew over.

This part has to deal with the preparation of the ingot and then how the silicon slices should be to get high yields when the chips are tested.

The process of refining the silicon crystal is one of the most important step for the progress we have talked about and let's see why.

The following image shows how in forty years has grown the diameter of the silicon slices.

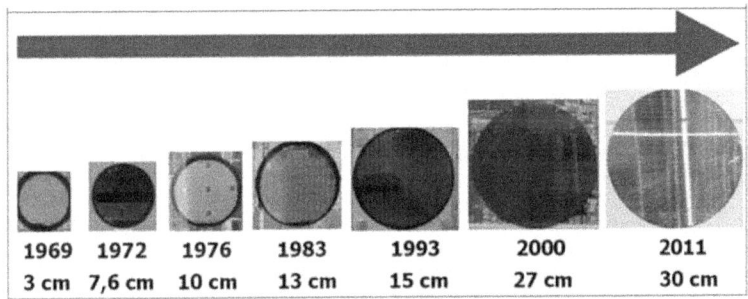

1969	1972	1976	1983	1993	2000	2011
3 cm	7,6 cm	10 cm	13 cm	15 cm	27 cm	30 cm

Wafer diameter increase over time

Why the ingots and the slices must be extremely pure and flawless?

If an impurity or defect of the ingot falls inside a chip its operation is completely compromised and must be discarded.

To avoid this issue, we must delete all the defects and impurities in the ingot before to slice it.

But it is impossible to delete them all, but over time there has been a continuous improvement.

And this improvement allowed this industry to increase the size of the ingots and so the size of each chip.

Can you imagine what kind of improvements had to be found to reduce the density of defects and impurities in the such a big ingots of today.

This technology of crystal purification is one of the greatest advancement of this industry.

In the next chapter we will look at the most important product that the invention of integrated circuits has brought to us thanks to all we have seen, a fundamental component for all the applications every day we deal with.

I am talking of the microprocessor that would deserve the Nobel Prize and that sooner or later will be assigned to its inventors.

The Birth of the Microprocessor

Up to this point we have been talking about integrated circuits containing a limited number of Transistors, like the integrated circuits used in computers of the sixties.

In these computers, the various single components, transistors, diodes, resistors, etc. were bonded to each other to create the digital units defined by acronyms such as RTL, DTL, and TTL, which constituted the basic logic elements of the computers.

These logical units were the first to be built as integrated circuits: it was like replacing a group of components with a single chip which occupied much less space and was less power hungry, thus realizing the objective to miniaturize those large electronic complexes that until then the industry churned out mainly for military uses.

With the increase in the number of Transistors contained in a single chip, it soon the industry succeeded in integrating an entire system. This very important leap was accomplished in 1971 at Intel.

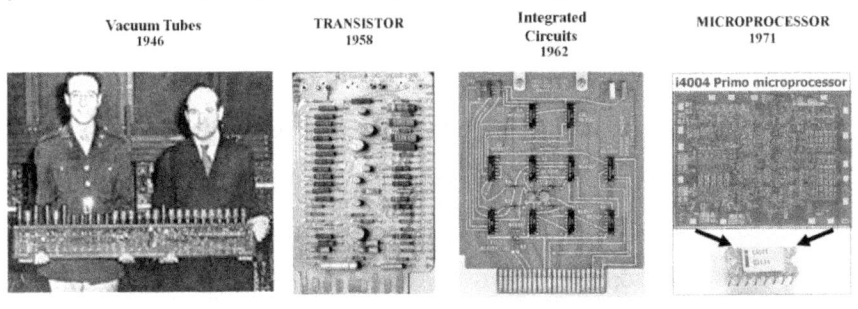

Development of digital computer electronic boards over time

With the microprocessor on a single chip we have a complete system, virtually the entire central part of a computer in a single chip.

This massive integration can be traced back, as we said, at the year 1971 with the creation of the first Intel 4004 microprocessor.

This chapter is devoted entirely to single-chip microprocessors, the most important chip of our times.

I must say I have been fortunate to be at Intel in the years 1970 and 1971, where a group of engineers had started working at the project to build a set of chips that on the market was called MCS4 and that included the 4004, the first microprocessor of the history.

On those occasion I had the pleasure of meeting Federico Faggin, Stan Mazor and Ted Hoff and some collaborator coming from the far east, among them Mathew Miau and Masushita Shima.

In a visit of mine in the year 1971 I learned that the Faggin's department headed by Las Vadasz had developed into a single chip virtually an entire computer intended for the Japanese customer Busicom and that all the work of the physical integration into a piece of silicon had been done by the Italian Faggin.

I spoke several times with the creators of the first microprocessor and I have met them also with our families.

According to my talks by then I can today confirm that both they and Intel didn't understand the importance of their work.

Even the creators had underestimated what was going on and didn't foresee its future and this is one of the reasons why Faggin left Intel in 1973; the company was not in favor of investing heavily in those products.

This fact is so true that in the second half of the seventies, when chips competitors became very aggressive, I received from Intel suggestions like to give away free of charge a microprocessor if I was running the risk of losing an order with many memory chips: the microprocessor was used as a bonus!

Even at customer site the situation wasn't easy either. Computer manufacturers didn't like to see a chip manufacturer to invade their own territory by producing in a single chip what practically was one key part of a computer they were used to design.

The difficult situations with clients at that time is described very well in a book whose Faggin is author.

In a paragraph with the very significant title "winds of war" he reports a meeting in Germany where he went to propose the microprocessor.

In that paragraph he says: "The worst meeting I had in Germany, has been with Nixdorf Computer, an important German manufacturer of small commercial computers, where we were almost ridiculed for the rudimentary architecture of our machine.

Some of these criticisms were certainly valid, but the level of hostility could only be justified by a more or less clear awareness that it was about to begin a war with semiconductor people. Well, thirty years after we know who is standing.

The first applications of the microprocessor were solicited by non-computer customers who had a problem to solve, and not time to waste by philosophical debates.

One of the first application won has been the spacecraft Pioneer 10, launched on March 2, 1972, which is the first object equipped with microprocessor to enter the asteroid belt.

In many other applications requiring low volumes the microprocessor, coupled with a software program, showed a much more flexible and economical solution, compared to the design of dedicated hardware". Up to this point how much Faggin reports of his experience.

This was the atmosphere in which we operated and certainly today, looking back, it may seem absurd, but it was true: the microprocessor was hated by many!

After many years it has been written much about the importance of microprocessors, both articles and books ... how things, at a distance of time change!

When I visited Intel in the 1991 I could see a photographic memory in its Museum, at the new headquarters in Santa Clara, building dedicated to Robert Noyce and located at 2200 Mission College Blvd.

I was there with my entire family (wife plus four sons) and I had the great honor of being escorted by the then CEO Dr. Gordon Moore.

Immediately to the right, at the entrance of the museum, I was able to photograph the section that remembered the event of the birth of the microprocessor, an event that would become the cornerstone of the Intel Empire today.

Santa Clara 1991 - Intel Museum – The microprocessor team

From left, Ted Hoff, Masushita Shima, Federico Faggin and Stan Mazor along with product images and descriptions provided a perfect representation of the reality I saw at Intel in 1972 and also the following years.

Shima played an important role in that adventure as it was the engineer named by Busicom, Intel customer, to collaborate on the realization of those chips that would be used at Busicom to produce computers for the consumer market. Intel then bought the project and transformed those chips into a kit called MCS4 and sold it to the open market.

The physical realization of the entire 4004 chip and therefore the integration into a single chip was made by Faggin to whom has to be credited the merit of the technological part of this creation.

In one of his books (published by Adnkronos in Italy, entitled "Faggin father of the intelligent chip") he describes with interesting details the patient manual work to realize that chip.

Today these chips are designed with the help of computers and robots, while the 4004 was entirely hand-drawn.

Below the photo of the first chip i4004 and its ceramic package.

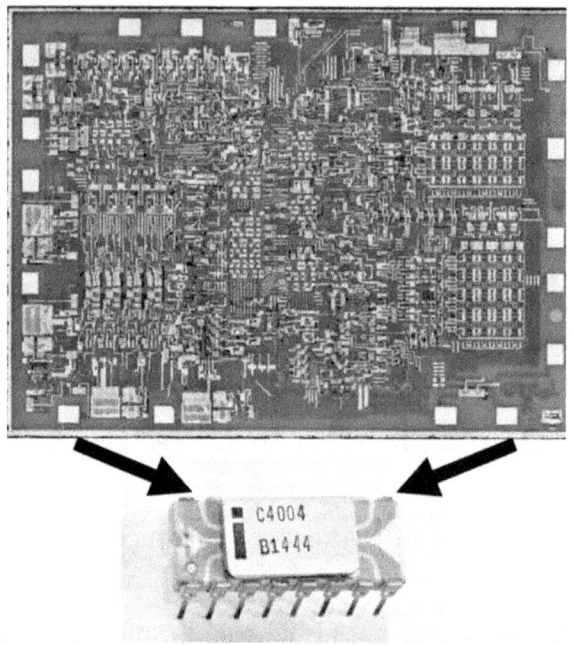

1971 – The 4004, the first Intel microprocessor and its 16-pin package

This chip, very complex at those times, was created starting from a drawing magnified and carving out a special sheet of mylar by hand.

Then the whole design was photographically scaled down to generate those masks that were used to produce the individual chips.

The 4004 chip was no bigger than a few millimeters on each side.

Here below the electrical characteristics of this 4 bit microprocessor, from the original document handed over to me by Intel then.

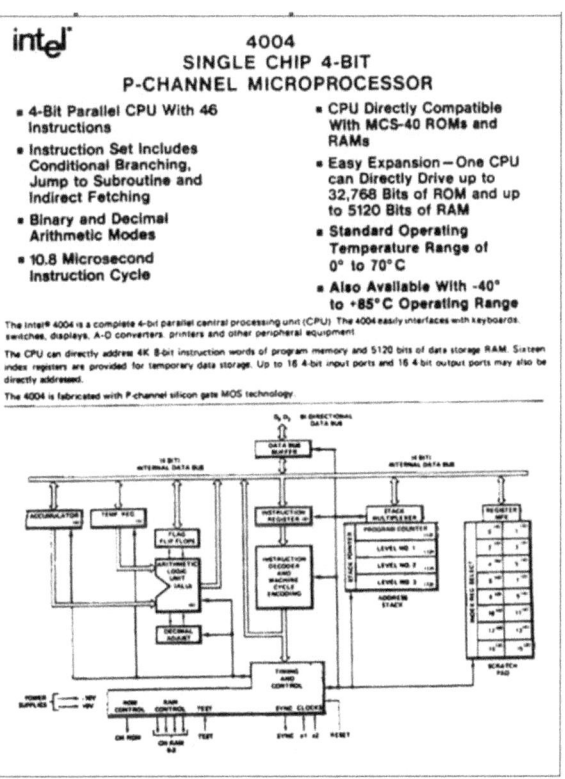

1971 – 4004 4-bit microprocessor data sheet

I had the chance to meet often Faggin during my travels in the USA and to follow the adventures of the other companies he founded after his release from Intel in 1973.

Said that, I must mention Zilog, the first company Faggin founded just after resigning from Intel, where he realized what became the most advanced microprocessor of the time, the Z80, and still used today in certain applications.

Palo Alto 1985 - convivial dinner in Palo Alto
with Federico Faggin and his wife Elvia

After the Zilog adventure, financed by Exxon, Faggin founded the Cygnet in 1982 with the truly avant-garde goal of realizing and selling integrated communication systems for the use of futuristic offices interconnected electronically.

It was the dream of eliminating the paper thanks to the integrated use of networks able to transmit and manage video, audio and data through appropriate software, precisely created by Cygnet.

This was thought by Faggin much before the advent of the commercial Internet and then using expensive private networks.

It was certainly a technological thinking ahead of his time and that in any case has sowed the seed for an integrated use of personal computers in business for the management of all the activities, an idea which will explode industrially in the 1990s.

Another interesting adventure of Faggin was that in the advanced world of artificial intelligence.

His idea, as a scientist of the logical processes that pushed him to the creation of the first microprocessor, was completed with the Synaptics, a company he founded to realize neural networks mimicking the processes of the human brain.

Talking with him I realized that his insight was to believe that wanting to really realize electronic structures able to duplicate some of the activities of our brain we should follow patterns very different than those hitherto conceived and that these roads were sued for building systems that are able to learn independently as children do.

These searches are still far from leading to concrete solutions, but who knows that in the future that seed that with Synaptics it has been planted will not flower sooner or later.

Anyway, waiting for big neural networks, Synaptics has introduced an element almost intelligent that today is part of a large number of portable equipment, I mean the touchpad, that smart mat that replaces the mouse and that is able to intelligently detect the position, velocity and pressure of our fingers.

Nowadays the "touch" technology is the most popular way to control computers and smartphones and we owe part of it to the research carried out at Synaptics, a fact little known and scarcely mentioned by our technological historiography.

I then met often also Stan Mazor, one of the three to whom is credited the realization of the first microprocessor.

Also with him I had interesting conversations about the development of advanced microchip and new computer aided design systems to further reduce the size of each Transistor in a chip.

Mazor also left Intel many years after Faggin and continued to cooperate in the industry's progress.

In a meeting with him during my trip to California in the 1980s he invited me to visit his new adventure at Synopsys.

At Synoptys they where manufacturing advanced Electronic Design Automation systems, hardware and software, simulators as well as transistor-level circuit simulators.

The simulators included development and debugging environments which assisted in the design of the logic for chips and computer systems.

Synopsys technology was at the heart of innovations on systems to improve the automation of the design and production of the ever more complex microchips.

1988 Mountain View. Stan Mazor's new venture and finding him heavily involved in EDA (Electronic Design Automation) ... could he do something different after having co-authored the invention of microprocessors?

The Importance of Microprocessors

The microprocessor is now a component in the common jargon and has assumed a central part in zillion applications.

In this chapter we will talk about its pervasiveness and how today we find it in almost every product we use, many times without us noticing it.

Small, almost invisible and silent manages many activities on our behalf.

We find it in washing machines, in car, in planes, in smartphones, in toys, in watches and even in clothes.

We find one or more microprocessors in all the products where it is convenient for us that they manage something in our place.

In other words, wherever intelligence is required, the microprocessor is used as an element that must make decisions based on the conditions set by the designer, processing it to decide what to do best.

Its function may be from the simplest decision like to command the washing cycles of a washing machine, to more complex decisions on how to take into account all the parameters of a plane in flight.

Its application may be the most simple decision, such as controlling the washing cycles of a washing machine up to more complex decisions, as to take into account all the parameters of an aircraft while flying.

The small microprocessor allows the distribution of intelligence in all equipment that we create and not just on computers, main use early in its history, but also in a wide variety of applications to make a product more effective and often cheaper.

We must then consider the exponential evolution that the same microprocessor has had in years and that have increased its capabilities, starting with the first one of 1971, a bit like the brain of living things, from the simplest organism to man.

Comparing a microprocessor with the central unit of living beings, that is, their brains, is still a daring comparison, but both central units have many similar functions: memory, decision unit, input and output units.

In the human brain each of these functions has grown with the evolution of the species while in the microprocessor they grow with the advancement of Technology. The microprocessor has evolved from the simple 4004 of the 1971 to the much more complex Intel i7, and more, in a time millions of times faster.

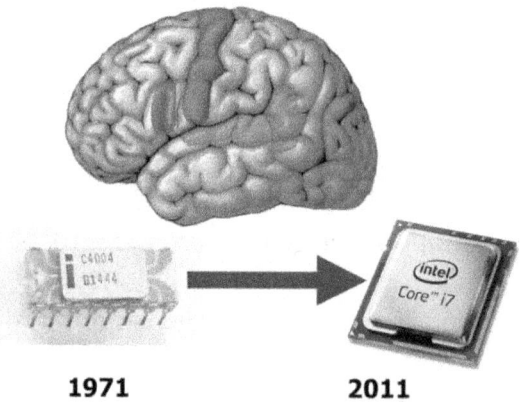

1971   **2011**

The microprocessor is the place where the computer processes data

We are still far away from being able to call the microprocessor with the designation of "electronic brains", if we want to equate it with the human one. Our brains are extremely complex, much more even of the most advanced microprocessor that technology is able to produce today.

However today, machines governed by microprocessor replace the man in many operations. Technology is still far from realizing all the functions at the same level as the human ones, but we cannot deny that at that level we are getting closer and quickly.

By making a parallel with the human brain and extending the Moore's law to a not very distant future, we will see later on what impressive predictions we can reach.

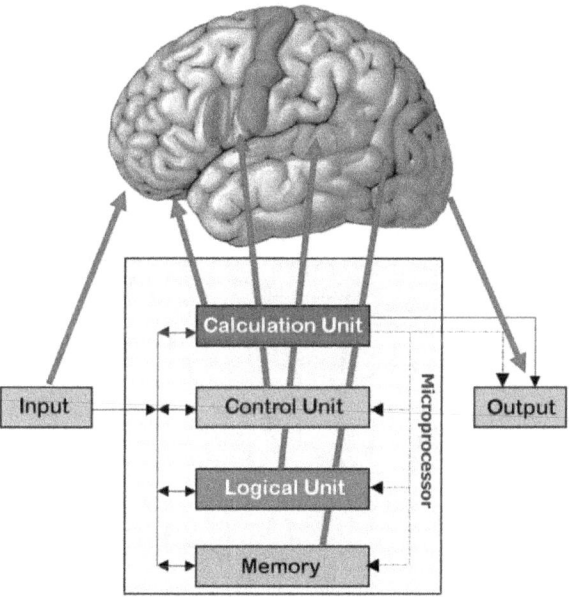

Also in the microprocessor we can identify areas of specialized activities as in the human brain

What has been illustrated here is an example of how a modest input action, looking towards something, once perceived by the human brain and elaborated produces an action such as to escape.

The process of evaluating what has been seen and to infer the appropriate reaction is far from simple. Requires a complex program that provides what to do in all the possible situations that occur in front of the eye, or in front at the webcam.

In many cases this is achievable electronically with the available microchips, providing only few images are analyzed.

Generalizing this matter will require much more computing power than we have today and above all, requires software like the one that only now the modern artificial intelligence (AI) is in the process to create.

According to what is happening today we can expect that in a couple of decades, engineers will be able to make three-dimensional chips with hundreds or thousands billions components inside and with the appropriate programs a microprocessor will be able to translate a language in real time as the brain of a great interpreter.

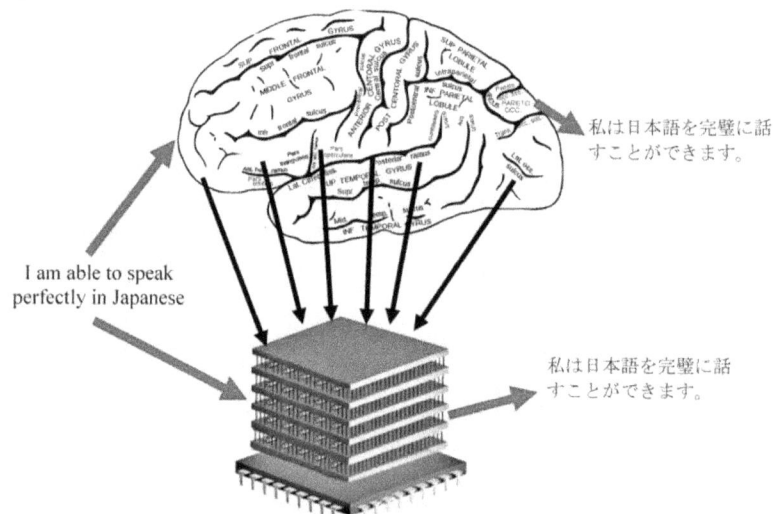

Future multi-dimensional hyper-chips that incorporate thousands of billions of components and software borrowed from the developments of artificial intelligence could be good enough at translating languages as a perfect bilingual interpreter

We will see in the chapter dedicated to the Moore's law, how by extending the validity of this law till the year 2050 we can draw conclusions that, together with the marvel possible, they can also leave us with some restlessness as in many science fiction movies.

Transistors Inside Computers

Before proceeding, we need to step back to our old Semimetallic Monsterling and to say a few words about what the Monsterling has meant to the world of computers.

From a giant valve system we have passed at Transistorized systems and finally, with the microprocessor, to a product for all: the Personal Computer.

It is obvious that the computers were the first application of a Transistor and then of the integrated circuits and how the Transistors have replaced the vacuum tubes used in large scale on the computers of the first generation.

The problems of the vacuum tubes were of three types: first their size, second their power consumption and third, the most critical, they failed in a short time.

As an example, in 1946 was built what is now considered the forefather of all electronic computers: the Eniac.

The Eniac performances were very modest, although to function employed 17,468 vacuum tubes, weighed 30 tons and occupied the space of a gym.

More, on average every hour one tube was failing and had to be replaced immediately.

That huge machine needed a small power plant to work given the large power consumption.

If a PC of today were made with the same vacuum tubes used by Eniac it would need many millions , there should be a large power plant to provide the power supply and would occupy the space of an average city, an accomplishment quite impossible.

We owe the solid state electronics the development of today's digital world; without the Transistor and what is followed, today the existence of all we know would not be possible.

*Eniac 1946 - The first computer weighed 30 tons, was employing
18,000 vacuum tubes and consumed more than 200 kwatt*

Thanks to the Transistor and its unique and malleable basic structure, a process was started to develop what we've seen, allowing to build highly complex equipment and at costs that have gone decreasing exponentially.

A single Transistor, was costing a few dollars in the 50s now, embedded in a chip, it costs millionths of a dollar: there is no other technology that has had a similar trend.

Talking about the computer and its history, it is necessary to mention IBM, the company that more than any other has marked the history of so-called intelligent machines and that in August 1981 began to market what would take the initials PC (Personal Computer) and which now sits on the desk of nearly all the officers of any company and on our desk too.

To describe the achievements of IBM and the many of its products would not suffice this whole book, so we will limit ourselves to mention the date of birth of the company and some of its product that have to do with our history.

IBM was born in 1886 with the name of TMC (Tabulating Machine Company), name that in 1911 was changed to IBM (International Business Machines).

As far as we are concerned there are two essential products that have profoundly marked the history of computers: IBM System/360 series, a very large systems sold since 1964 and the personal computer introduced in 1981.

The first is a gigantic electronic machine that makes extensive use of Transistors and integrated circuits; with its enormous computing capability provided the industry, the research and the military with the most formidable tool to elaborate the complex operations that in the sixties were needed both for the military and space race and for the sophisticated scientific analyses born at that time.

The IBM System/360 was created and sold in a variety of configurations, from the smallest, the model 20, to more complex ones called models 195, the latter with a 4 megabyte RAM and with a CPU execution cycles time of 50 nanoseconds.

In other words, while today a personal computer has an average 4,000 megabytes of RAM and speed of execution of the instructions of the same order of magnitude, the IBM model 195 was costing several million dollars against 500 dollars of an average personal computer of today.

*1964 - IBM 360 series. The new era of industrial computers
began and was a great commercial success*

Not to mention the complexity to maneuver it: it needed highly skilled and expensive operators; just look at the console to govern it, that is, what today is the keyboard in a PC, with all those switches and knobs that would scare even the most daring person among the authors of science fiction.

The complex control desk of an IBM 360

As mass storage, instead of the current hard drive, huge tape memories were used to search the information and they had to play the tape back and forth, with all the resulting slowness.

Loading data and programs was done through a great deal of punched cards that were written by special perforating machines and other machines, reading the holes, would move the data into the main memory.

The input data of the first computers were punched on special cards

Investment in this series were enormous; it appears that IBM has spent more than 10 billion dollars of that time, billions that made computers to take a huge step and it must be said that with this series IBM introduced innovative devices and software of the highest level that the world of personal computers has adopted, as the disk storage system, hard drives in today's terminology.

The second milestone of IBM has been to bring the PC to the professional market thanks to its prestige. Apple and Commodore at that time were mainly in the hobby world.

August 1981 – the IBM-PC was presented to the public

The serious IBM did with this product a masterful marketing operation, still remembered for his courage and effectiveness.

In fact the world was accustomed at seeing the IBM as something unreachable for the most, the forge of wonderful machines, powerful and ... overpriced.

To launch the personal computer to all the people, as they wanted, required the use of a testimonial certainly not traditional at IBM and someone decided to create ads for the PC with a popular and comical figure like Charlie Chaplin, combination that had great success and sales skyrocketed at levels dozens of times greater than any more rosy forecast.

1982 - IBM Ad to publicize how much a PC could handle

To conclude this chapter on the Transistor inside personal computers we must mention Apple Computer that had preceded IBM by 5-years.

The two companies had very different computers: the IBM was based on an Intel 8088, a 16-bit CPU with multiplexed outputs to stay in an 18 pin package. Apple was based on a 6502 CPU produced by MOS Technology, then a very popular product.

When the IBM PC was released on the market, the best-selling personal in the world was the Apple 2 which, among other things, I was personally selling in Italy as well as using it with the excellent VisiCalc program from VisiCorp, the first spreadsheet in the world.

Steve Jobs did not appeared frightened by the entry of IBM in its market, even publicly gave them the welcome.

Welcome, IBM.
Seriously.

Welcome to the most exciting and important marketplace since the computer revolution began 35 years ago.

And congratulations on your first personal computer.

Putting real computer power in the hands of the individual is already improving the way people work, think, learn, communicate and spend their leisure hours.

Computer literacy is fast becoming as fundamental a skill as reading or writing.

When we invented the first personal computer system, we estimated that over 140,000,000 people worldwide could justify the purchase of one, if only they understood its benefits.

Next year alone, we project that well over 1,000,000 will come to that understanding. Over the next decade, the growth of the personal computer will continue in logarithmic leaps.

We look forward to responsible competition in the massive effort to distribute this American technology to the world. And we appreciate the magnitude of your commitment.

Because what we are doing is increasing social capital by enhancing individual productivity.

Welcome to the task. **apple**

Transistor and Consumer World

At the beginning the industries manufacturing Transistors and integrated circuits addressed their interests mainly to the market of large computers, military and space because their products could satisfy well the requirements of size, reliability and power consumption.

On the other hand, from the economic stand point of view, producing a chip is a bit like printing books: the more the copies are less each copy costs and this is one of the reason it has brought this component to cost less and less by producing it by the millions and those markets were absorbing Transistors and integrated circuits in very large volume.

In fact when the Transistor came to light didn't exist any other areas beyond that of big computers where to use identical quantities of the same component: a large computer in fact contained immense amounts of repetitive digital logic elements made by Transistors and for this reason the computer market was flooded with Transistors.

After a while, as we have seen, the digital logic units have been integrated, a natural technological step. Production cost of computers went down and their performances started to skyrocket.

But our Transistor soon has put his hands also in products other than computers, invading also the consumer market.

Radio and television sets manufacturers became aware of the existence of the Transistor and soon found a way to improve their products by using them.

In the mid-1950s were indeed available the first Transistorized radios and it is precisely in the area of audio reproduction via radio or in the form of amplifiers, more or less powerful, that small portable electronic products made their entry into homes, more than twenty years before the personal computer.

And here especially Japanese companies have been shown to be able to win the consumer market by investing a lot in Transistorized items for the joy of the public.

These Japanese products invaded in a short time the rest of the world for their reliability and relatively low cost.

I remember the Transistor radios that could stand in the palm of our hand like the Sony TR620 and I remember football fans in the stadiums holding close to their ears this radio to hear the commentaries of other matches.

Sony TR620 - Small Transistor radio that invaded the world in the 1950s

Japan then retained to the present day its worldwide leadership towards consumer electronics by including cameras, television sets and everything that has to do with sound and images.

During that time I liked to do something practical in the area of Transistorized devices by building and publishing radio and Hi Fi amplifiers projects that were based on Transistors made in Europe by Philips Semiconductors and by the Italian company SGS (now STmicroelettronics).

I found an historical photo of a circuit I built by myself while I was a student that contains about 30 components. In the following picture I compare my circuit to a circuit extracted from a modern laptop that contains at least 10 million components: a nice improvement over my old modest construction!

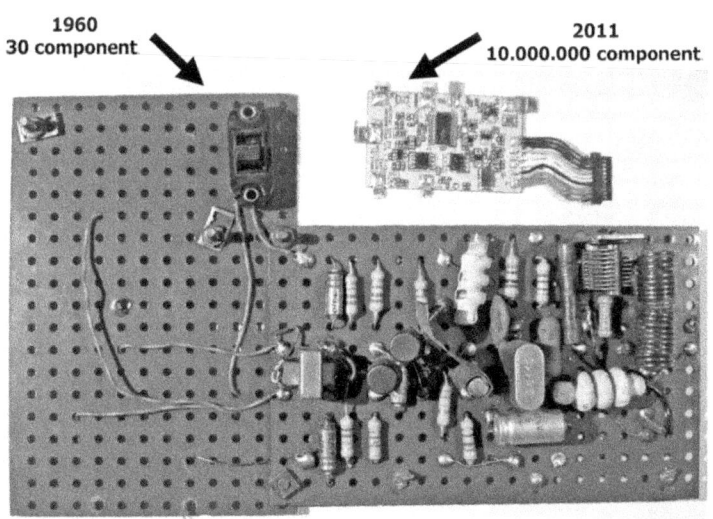

A 1962 self-built transmitter with three Transistors and the small internal circuit of a modern laptop computer with 10 million Transistors

I still remember with some nostalgia the names of those old Transistors used: OC47, 2G222, OC71, OC72, OC16, 2G240, 2N34,

CK722 with whom I was assembling my projects like portable radios, transceivers, amplifiers by night between the preparation of exams.

The foundation then for using Transistor circuits also for use in consumer devices were thrown and their future will prove even more fruitful than space and military applications.

We have to wait until the 1980s to see large volumes of digital equipments in the public market and once again all is due to the Transistor and the Moore's law.

The portability of electronic devices with transistors was the great magnet of the public in the fifties, sixties and seventies.

You could see boys and girls, myself too, to turn around with cumbersome headphones connected to the popular Walkmen, couples on tour with player and cassettes, people in the car with heavy televisions that of portability had just the handle.

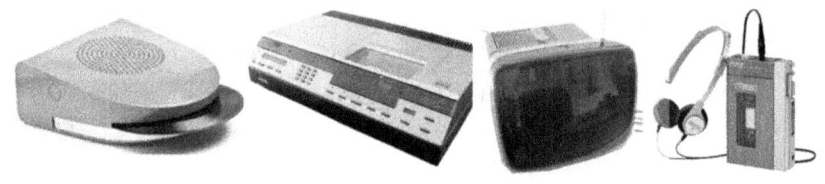

In the 1960s and 1970s, portability became a must ... from Japan

Complexity vs. Intelligence

Since equipment using a growing density of integration is becoming more and more intelligent, we have to ask ourselves a delicate philosophical question: "can the intelligence of these devices be considered as we humans mean as intelligence?

The answer to this question is part of the endless discussions technicians, philosophers, metaphysical, religious have been asking themselves for at least a century.

Although by sure the purpose of this book is not that to deepen such a very delicate topic, however we can add some modest consideration, also because this is the needed premise to a next chapter where we'll cover the Moore's law and its future consequence.

Not to mention the enormous importance that these considerations interest the advent of the future microprocessors that we will study in the next chapter.

We know that the logic of a computer is based on elements that have only two states, namely the one and zero, and this element is called bit. The bit is a unit that can count with just those two values while our hands, as an example, can count up to ten having ten fingers.

If the human being, instead of having ten fingers, had been created with only two fingers, probably we would be using the binary logic from the night of the times instead of counting on base ten ... or, with only two fingers we would not have been able to achieve anything and today we would not even be here to discuss about this question.

Apart from the philosophical aspect of this question, let's see how today we could define the artificial intelligence as shown by our computers and the human intelligence, limiting ourselves to a technical analysis, just to understand what we are going to say later.

Let us first see how physically operate the two contenders, the computer and the brain, and let's compare them.

After having put together and linked a number of Transistors and inserted input equipment like keyboard, mouse, webcam, and output devices such as monitors and printers, we have created a system that can operate automatically according to a program, which we call software, and that the various Transistors, opening and closing as switches, perform faithfully: we call this system a "computer".

Even our brain has senses at the entrance like sight, hearing, tact, smell, taste that elaborated by the brain provoke outgoing reactions such as walking, tasting, listening, touching and an infinity of other actions.

So the two systems, the artificial and the human, enjoy the same ability to provide output actions based on input data governed by appropriate programs. Frankly speaking   the programs that humans use are a little bit more complex!

To someone this comparison may seem a bit irreverent, but there's no doubt that from the perspective of only principle the humans have been   able to accomplish something that reacts intelligently to input data exactly how does a computer.

Of course it must be argued that in a computer the program must be created by humans, otherwise the computer would not do anything and here is the first, far from being negligible, big difference between the two.

Clear that then nowadays, no matter how powerful and complex a machine is, we are still a long way from being able to mimic many human functions, but saying "nowadays" we introduce a variable, the time, which can make us to think of a possible future in which the two contenders will come to be equivalent and we will see it in a next chapte.

Our considerations, more technical than philosophical, cannot run out here, so let's take a little step forward and see what activity of a human brain can be compared with the one of a computer, always with the precautions of the case.

INPUT DATA   OUTPUT ACTION

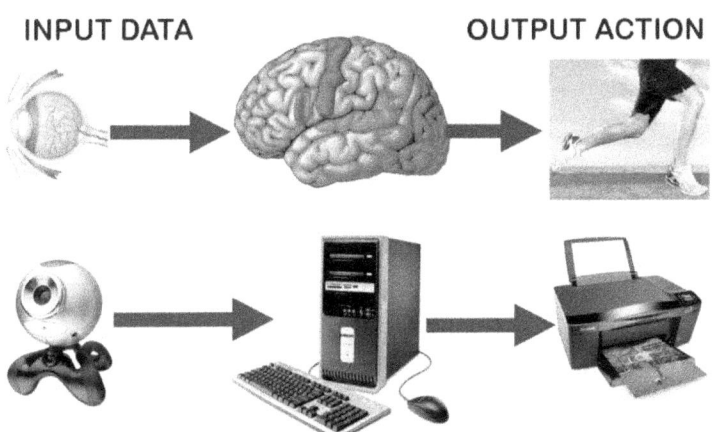

Like humans, even the computer has ways to perceive the external world and outbound actions according to a program

Yes, in this case a comparison is possible: the computer receives data from a webcam and processes them with a program specially loaded on the computer; the computer then is able to operate a printer and to give us a nice picture of what the webcam has seen.

Similarly the human eyes see something and if this something is a danger, the brain is prepared to understand it and if the danger is approaching the brain communicates to the lower part of the body to escape quickly.

A keen reader might notice that something else is controlling a printer and to be able to distinguish between an image of danger and another image that doesn't show any danger at all.

And here is the very substantial difference between the machine and us: we are able to interpret what surrounds us in a way that, to date, machines are not able to do. But a question arises spontaneously: will one day the computers succeed in doing that?

This is the key point: if the complexity lies at the base of our ability then perhaps, in a few decades, the machines will reach our complexity and they should be able to simulate us in a perfect way.

And as we will see in the following chapters, there is a near certainty that this will happen in three or four decades and at that time we would be entering an era in which science fiction robots would become a reality almost human like.

However, the fundamental doubt remains if there is something more than just complexity in the human mind, as the author of these pages believes, if so the machines will be able to overcome some human capabilities, as already in many applications today, but they could never supplant us in our entirety ... at least this is hoped.

After this brief chat about artificial intelligence it is the case of starting to treat the Moore's law that we will evaluate in the next chapter, a chapter that has to do with the concept of complexity.

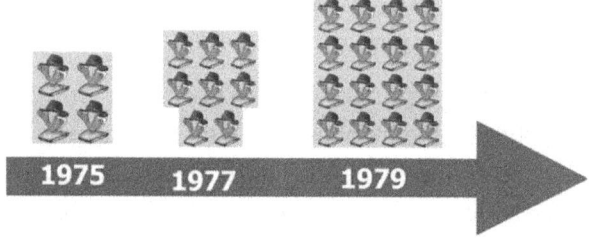

Moore's law: in a chip Transistors contained double about every two years

This very simple law contains enormous consequences if it gets extended overo many years, in fact implies a process that grows exponentially and that quickly reaches enormous values.

I remember at this regard a famous tale of a King who, for gratitude to his mathematician, asked him what he wanted as compensation for his performance and the mathematician responded in a simple and seemingly trivial way, saying: "I want so many grains of wheat as they result by putting on a chess board a grain on the first box and then doubling for each next box until the last box".

The king was surprised at this seemingly modest request and asked his other scientists to calculate it and immediately to satisfy that request.

These scientists immediately began to reckon by doubling each box as required and realized soon that the result would have been so great that it would not suffice all the crops of the entire planet to satisfy the request of the King.

That's exactly the point hidden in the simple enunciation of Moore's law, the consequences are unimaginable if the law were to be extended again for many years, as is, by sure, going to happen.

Just to satisfy the reader's curiosity, if the scientists of the King had reckoned righteous would find as a result a number of approximately twenty followed by eighteen zeros, i.e. something like 20 billion billions grains and if each grain weighs one gram that outcome would result in something like twenty thousand billion tons of grain and to carry them with large ships of 10,000 tons each, it would take two billions ships ... clearly impossible and ...... well crafty and cunning that subject!

From the mathematical standpoint of view we are facing a process that grows exponentially and that, as we have seen, leads to quickly reach enormous dimensions, values hard to find in other human activities.

This simple prediction, which then in the 1970 a journalist will call Moore's Law, is in fact part of an article that Dr. Moore published in the 1965 in the Electronics magazine and that the author of this book, subscriber to that magazine, read without understanding minimally the implications.

It should be noted that Moore, co-founder in 1968 along with Dr. Robert Noyce of Intel, when published his article was in charge at the Fairchild Semiconductor laboratories and it is therefore fair to trace back to this company the environment in which Moore came to this insight.

What exactly did Dr. Moore wrote in this article? That in the visible future (he foresaw until 1970) every 18 months the semiconductor industry would be able to produce at equal cost integrated circuits containing twice the number of Transistors.

In 1991 I went to visit the new Intel building in Santa Clara at 2200 Mission College Boulevard with my whole family.

In that occasion, as I described in a previous chapter, Dr. Moore kindly escorted us on a tour at the factory and at the end we went to the cafeteria.

There he confirmed to us that his thought in 1965 was that this would have been possible only for a few years and then added that he could not imagine that his prediction could last till the moment we were conversing and I should add now, up to the present day and by sure into the future too.

1991 Santa Clara – I visited with my whole family the new Intel factory and I met there again Dr. Gordon Moore, with whom we remembered the epic seventies

That factory was truly a wonder, much bigger than the last one I had visited a few years before and this building was also constructed earthquake-proof.

Moore took me to see the foundations of the factory whose pillars of reinforced concrete at a certain height were cut by a huge rubber cushion which could withstand vibrations of even the most disastrous earthquake.

This factory is the heart of Intel and if production stops there for a long time all those silicon slices that feed other Intel factories around the world would stop eighty percent of the world's personal computer production.

1991 Santa Clara – Factory tour with my whole family

After that, hooded with the usual protective head caps we entered the white chambers, that is, those environments where everything must be perfectly clean for the delicate operations that take place there; we continued visiting and admiring the interesting operations to produce the various microchips we described earlier.

I could see the size of the silicon slices, which the robots handled with dexterity moving from one implant to another and I could observe those powerful masking plants that were followed by furnaces for the various stages of diffusion: it seemed to me to be on another planet.

Modern white room

At the end we could not miss a short breakfast with Dr. Moore at the cafeteria of the factory, where we exchanged some chatter on the perspectives of the ICS.

Above all we could remember the pioneering times of the early 1970s and when we were running on the highway Milan-Turin to visit Italian customers, travels that he confirmed to me to have done with terror for my "crazy Italian way of driving".

1991 Santa Clara – Stop at the cafeteria of the factory where all of us were listening to Dr. Moore's interesting words

Once out from the factory I took a photo of the brand new entrance of this huge complex, still partially under construction and the dedication to Dr. Robert Noyce that passed away the previous year.

1991 Santa Clara, view of the entrance of the new Intel factory

Returning to the conversation with Dr. Moore in which I said that he himself did not foresee with his article what would have happened looking at far ahead his time, I should tell my reader that the production techniques have continued to hone as no one could foresee and his prediction, I believe, will maintain validity for decades ahead ... I am sure.

In a nutshell, it seems that this huge growth in the number of Transistors in a single microchip can grow to infinity, which obviously cannot be, exactly as could not be found all that grain required by the king's subject with his cunning question.

Anyhow it is not excluded that one day we could manufacture microchips containing even hundreds of billions of Transistors and perhaps more, of course if Moore's law will hold for a few decades more.

Let's try to understand on what basis and what concrete facts has inspired Moore by writing his famous article and why the whole thing is lasted till today and how it all depends on the particular way in which the integrated circuits are manufactured.

Let's start by analyzing the original chart published in 1965, a graph hand drawn by Dr. Moore.

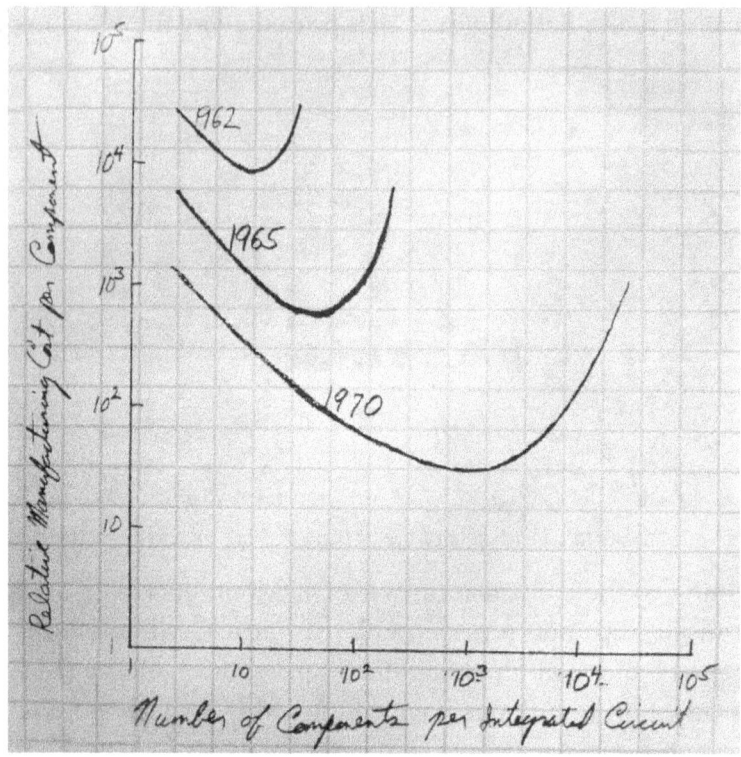

1965 - Electronics magazine – the original graph of Moore's law

The graph tells us that in 1962 the minimum cost per Transistor was obtained in production with an integrated circuit that integrated little more than ten Transistors.

In 1965 the graph moves this minimum forward to chip that integrate some hundred Transistors and the same graph predicted that in 1970 this minimum could be around some thousand Transistors.

If today we look back over the years and we put in a chart what happened from 1962 to 2005 we get the following chart published by Intel and that pretty much covers the provisions of Moore's law until 2005.

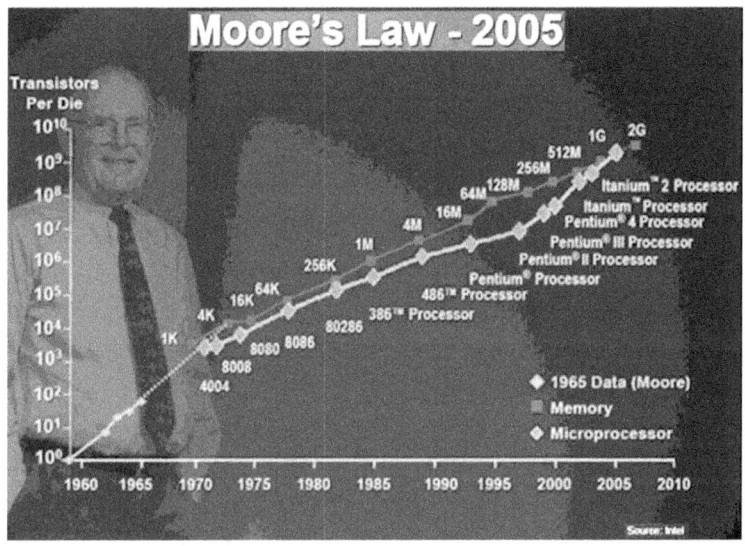

2005 – as Moore's law actually occurred

At this point, Moore's law is demonstrated for more than three decades and we can see the implications for all that is happening around us today.

Never in the history of technology, at any angle of human development, there has been another activity where something has developed in such a short time and so fast.

Somebody, somewhat humorously, has made a comparison stating that if happened the same in the automotive industry, today a car should cost less than one dollar, run to over 100,000 miles per hour and consuming less than one cubic inch of gasoline per mile, and the comparison is definitely at fault than the reality of integrated circuits.

This comparison is a bit pulled for the difference between the world of mechanics and electronics, but it gives the idea of what kind human event we are facing now, and how the Transistor and the following technology of silicon integration is moving the whole of humanity towards something that develops under our eyes at an incredible speed and how the whole matter is an absolute novelty of the history of our civilization.

Let us deepen the reasons that make this development possible and let's try to grasp the intimate technical reasons in order to extend the effects to a visible future and let's see the incredible consequences.

We should remember what we saw in the previous chapters, especially when we described the production method at Fairchild invented by Dr. Robert Noyce in 1958 and how that method allowed to "print" entire electronic circuits in a chip with many components, all connected to each other, realizing what we have up to here called integrated circuit.

In practice the microchip is nothing more than a tiny rectangular of silicon crystal, mostly not bigger than a fingernail (you have seen photos on a razor blade and a pencil) and that is clipped from a thin slice of silicon obtained by slicing, just like a salami, a cylindrical ingot of purified silicon crystal, slice on which are accomplished all those operations that serve to realize those many sandwiches which are the integrated circuits.

These salami slices can reach a considerable size; today even more than fifteen inches in diameter, so on each slice there are many of these tiny rectangles destined to become microchip, even hundreds.

Each microchip in turn contains a large number of Transistors to form a complete working system, number that obviously depends on the size of the chip and the size of each Transistor embedded in it.

It is therefore clear that when the size of the Transistor is diminished and the size of the silicon rectangle grows, the microchip will always contain more Transistors.

You will have already understood that if we increase the chip surface and simultaneously decrease the size of the Transistor by appropriate techniques, our integrated circuit may contain an ever-increasing number of Transistors and thus it becomes an increasingly complex electronic system.

The natural question that arises is how much you can increase the size of the chip and how much you can decrease the dimension of each Transistor to poke them inside the chip even hundreds of billions or more!

But things are not that simple: both operations have limitations and to assess these limits for the future was the genius of the intuition of Dr. Moore.

As a great connoisseur of the production techniques, not separated from a deep knowledge of semiconductor physics, Dr. Moore evaluated, with quantitative precision, what could be the progress of the industry over time in order to obtain how much his chart showed.

In other words, he envisioned how much each year the industry would have succeeded in enlarging the chip and at the same time how much could shrink the Transistor.

In fact, whenever there are precise technological limits beyond which the manufacturers cannot go and these limits require time to be exceeded, this timing has been magnificently predicted by Moore's chart.

Going ahead with our explanation, let's see from where these limits come from and what the engineers have to do to overcome them and thus produce more and more complex integrated circuits and at the same time reduce the costs of the individual functions produced.

First we should consider our little rectangle of silicon obtained by cropping the silicon slice and we should ask if there is a limit to its maximum size beyond which we cannot go.

It should be known that if a microchip incorporates an impurity or defect, as we have seen in the chapter where we have talked about production problems, this microchip will not work properly because one or more Transistors contained in the chip will fall right above those impurities that will prevent it from functioning.

This defect will compromise the whole integrated circuit which then must be discarded once manufactured.

Obvious that the ingot should be cleansed at best, but as far as one will purify it, some impurity, however small, will always remain, and then the larger will be the size of the chip to be cropped the higher will be the likelihood of incorporating an impurity.

If the distribution of impurity in the ingot and the size of the microchips are such that the probability of incorporating an impurity in each microchip is of one hundred percent, then no microchip produced will work.

We must therefore reduce this size to the point where only a few microchips for each slice will incorporate these impurities, that is until the spread of microchips containing a defect will be statistically lower than a certain value tolerable costwise.

Moreover, as one purifies a crystal ingot with the sophisticated techniques used today and whatever one does to prevent defects, these operations have natural limits and even if the industry pushes ever forward, improving over time their own means, the density of defects and impurities in the crystal will never be zero. Today we are with impurities of the order of a part per million.

Concluding, Dr. Moore knew that at his time the dimensional limit for a chip to not include with high probability an impurity could be, say, around a square inch, while proceeding over time, he could predict that after six years this chip could reach a working size of two square inches.

Today, with the purities reached, even chips of several square inches can be easily produced.

Once seen the first dimensional limit to the growth of the microchip, determined by the density of impurities and defects of the crystal of origin, we should see the second limit to the complexity of the circuit containable in a given microchip.

We know that every single Transistors contained in the microchip has its own physical size, whereby, given a certain microchip, if we are able to reduce the size of each Transistor by half, the resulting chips will be able to contain four times more Transistors.

But just as we have been looking to reduce the number of defects of the crystal and we found its limit, also to reduce the size of the Transistor has precise limits determined by the photolithographic technology available.

That is, each Transistor is fabricated using techniques borrowed from the world of photography and the size of Transistors on-chip drawable has its limit due to the ability of the camera that photographs the layout of the microchip.

Namely its capacity in terms of definition: the more the Transistor is small greater should be the definition that the machine must manage

This industrial capacity is not infinite and grows with time as for digital cameras that at the beginning had a megapixel sensor elements and today they largely exceed the twenty megapixels.

The semiconductor industry, driven by the pressure of competition, has been so good to pass in almost fifty years from chips not larger than a square centimeter and containing a few dozen of components to chips of over twenty square centimeters and containing over one billion of Transistors

And on the same time the key dimension of the Transistor, the channel width, as decreased from few micron to 16 nanometers,

For example an Intel Pentium CPU dated 1992 had this dimension equal to 3 tenths of a micron (300 nanometers) while the latest multi-core CPU are realized with gate equal to 22 thousandths of a micron (i.e. 22 nanometers) … and recently even less.

It has been already announced the next generation smaller than 22 nanometers, so to speak, it is the length of a few tens of silicon atoms aligned. The industry is already thinking about the next generations of 16 nanometers and 11 nanometers …. equivalent to a few atoms!

For the completeness of this fundamental chapter, I must remind that there are other limits in the realization of increasingly complex chips and among these a very stringent one is the dissipation of heat.

More the chip is compact and fast and the more heat it generates and exceeded a certain threshold of heat the chip self-destructs when switched on.

And this threshold was reached in recent years, for which reason the industries have been forced to stop increasing the speed of the microchips.

This obstacle was circumvented by duplicating the devices on the same chip and decreasing the speed of each device, thereby increasing performance but without overheating the resulting chip.

These so-called multi-core CPUs came to light, where in the same chip are packed two, four, eight, and who knows how many more CPUs, all equal to each other and working in parallel.

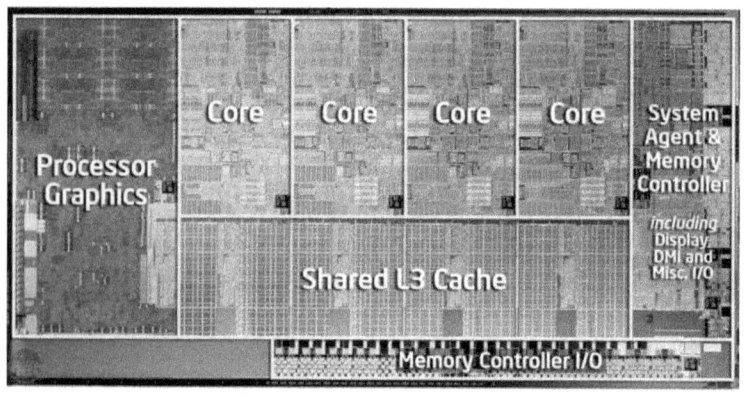

Modern chip that realizes a multi-core microprocessor (CPU)

There is another effect caused by Moore's law that has had incredible consequences in the world of semiconductor companies in the period that lies halfway between the late 1960s and early 1970s, the consequences that the historical literature seems to neglect.

I refer to the fact that the Moore's graph makes it clear how the highest density at the lowest cost of a chip moves very quickly with the time.

It is then clear that the competitiveness between companies had to move in the sense of arriving before the others to integrate the maximum number of components at the minimum cost and then to develop the techniques to achieve this goal and so to win the competition on the market.

So one could see, south to San Francisco, in what is now called Silicon Valley, a flourishing of large and small semiconductor companies, almost all founded by industry technicians, spin off from other similar companies, and that were fighting at the sound of miniaturizations and each one looking for their own market niche, trying to advance before the others on the Moore's curve.

And we can say without a doubt that in general and also today the companies that survive are along this curve!

I can say that I have seen all this with my eyes and every time I landed from the plane in San Francisco and I was doing my lap in the valley between Palo Alto and San Josè I could find new companies with incredible market objectives.

Just to name a few, in addition to Intel, National Semiconductor, AMD, Electronic Arrays, Zilog and many others and some still exist, while many have disappeared or were swallowed up by bigger companies and I remember that time as truly exciting.

There was an incredible euphoria based on the certainty of being able to stay in front of others and to find room in an immense market and to easily get source of money by the venture capitalists.

It is from that hotbed of ideas and practical achievements, many adventurous, which then were born those markets that today reach billions of people like personal computers and Smartphones; it is that enthusiastic environment that guys totally unknown and devoid of any industrial background such as Bill Gates and Steve Jobs could throw bravely the seed of what will become at our days the world's largest companies.

And the miniaturization and all those races are still there today, but in dimensions then unimaginable. Now we are at over 22 nanometers, and we do not see yet the end of this path, end that can only be decreed by the attainment of the physical limit of Moore's Law ... assuming that this limit exists, limit that we will analyze soon.

In my first 1969 book that I published while student I had reported a graph comparing the electronic miniaturization known at that time with the human brain, a graph that I report here below exactly as it was:

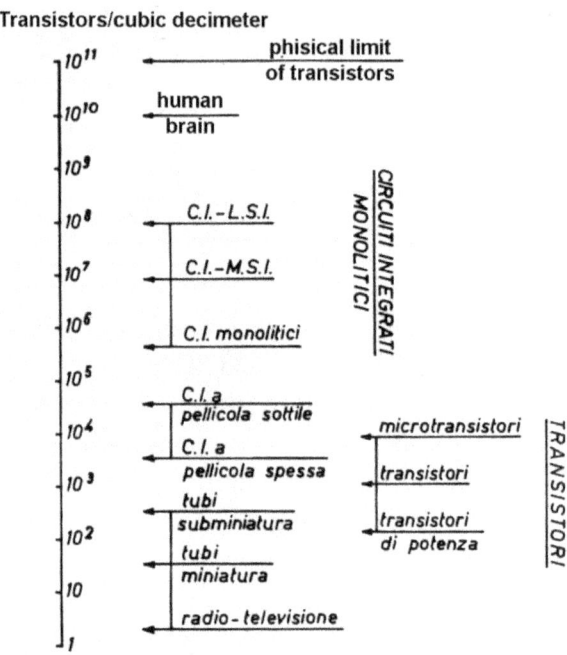

1969 – Scale of complexity as it was in my first book

It is interesting to note that, with some changes that we will see, this chart is still relevant and, most importantly, how the unthinkable may happen including to overcome the density of the human brain.

In the next chapter we will screen Moore's law in the future and we'll see what we can expect.

Moore's Law and the Year 2050

At this point we must return to the beginning of this book, exactly to the first image I showed to my 11-year-old grandson to solicit his interest and that I obtained as a result a complete failure.

Below that image.

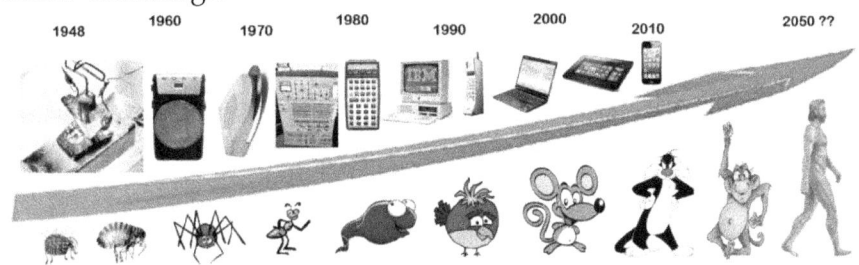

Evolution of equipments produced in 60 years thank to the Transistor; from a small portable radio of 1956 to the modern smartphone with its billions of transistors. Comparison with the evolution of nature over millions of years

After reading up to this point and then with the knowledge acquired I can explain this image but also project it into the future in the future.

Recalling the story of the King and his mathematician, we learned that by doubling the grains of corn on each of the 64 boxes of a chessboard we got such a huge quantity of twenty billion billions grains, namely a quantity of corn that our Earth cannot produce even considering all collected from when mankind has learned to cultivate.

With that type of exercise we are faced with an exponential growth, that is, a number that in mathematics is expressed by two raised to the 64th power.

Now, even with the integrated circuits we have a similar growth where at every box of the board we have to replace two years, at least this what Moore's law says and what the history has proven to be true for the fifty years elapsed until now.

Let's extend the reasoning by doing the following consideration.

Integrated circuits, or microchips as you can call them, have existed for about fifty years and we know that the first 4004 microprocessor was founded in 1971 with approximately two thousand transistors in it. To simplify our calculations, we can start from the previous year, i.e. since 1970, and we can assume that in that year the technology was able to integrate one thousand transistors on a single chip.

With this assumption, and having verified the validity of Moore's law until the date of 2012, we can reasonably believe that we can extend it in the future, the doubt being until when.

Forty-two years have passed since then, namely twenty-one-year periods, and if we use the method of the grains of wheat and a thousand Transistor at the beginning and doubling every two years, in the year 2012 Moore's law says that we should be able to integrate about two billion Transistors and this is really the best that the industry has come to make that year, as it turned out.

Thus, this extraordinary law of doubling every two years has hitherto proved absolutely correct and the question that arises spontaneously is whether and how it will continue to maintain this rhythm.

In this chapter we intend to attempt some exploration and to do a few thoughts on the future, knowing well that for several times any conjecture about the future of Moore's law proved false in a pessimistic sense.

Even the same Moore has been foreseen an end for it many years ago, but then new techniques and formidable industrial progress managed to overcome obstacles when these obstacles presented themselves.

Today the best experts predict an end very close because they feel that it is not conceivable a Transistor so small that approach the size of a few atoms.

The most optimistic extend the law up to the year 2024, when however, if Moore's law remains valid, we will have to be able to integrate at least thirty-two billion Transistors into a chip.

So with that doubling, at thirty-two billion Transistors, according to someone, the industry should stop: but it probably won't be so and let's see why.

It remains to understand what can be done with similar levels of complexity: in fact, the higher the number of Transistors the more complex are the functions that the microchip can perform.

In fact applications, such as voice recognition, real-time language translation, handwriting recognition are at hand now.

Surely it will be a surprise what we will be able to do with a chip containing thirty-two billion Transistors, bearing in mind that, for example, in a smartphone of today are already contained hundreds of millions of Transistors and it is a tool with wonderful capabilities.

Here the imagination can really indulge; probably you will be able to talk to an appliance in English and to get as output a perfect voice in Chinese or German.

Certainly in the scientific field, many of the calculations and analyses that today require hours or days at that time will take place in minutes or even seconds.

The automation of vehicles used every day, washing machines, TVs will reach incredible levels ... they will practically be able to perform a sea of functions at the mere snapping of the fingers and then, the more users will ask, the more industries are going to try to produce them.

Back to the many scientists claiming correctly that there is a physical limit to the advance of Moore's law, I have something to say.

At this regard I can't forget that I has been reading for thirty years comments about the close end of the Moore's law and it never proved true.

I remember when the technology reached the micron as Transistor size and was said the physical limit would be reached soon, around two or three tenths of a micron.

Then, that limit was overcome by innovative and unpredictable solutions. In fact, it was soon reached the tenth of micron, which was still thought that in a few years, and with half of that size, we would have stopped, but new techniques and new inventions intervened again to overcome it.

But then, if it's already so hard to make logical considerations for the year 2020 or 2024 what, on earth, can we expect, say, in the year 2050?

Here we enter a bit in science fiction and in fact, rereading the predictions made in the past we will notice how no one has ever managed to predict what then happened, despite often those predictions were supported by sound scientific arguments at the time deemed reasonable.

Sure, Giulio Verne at the end of the nineteenth century wrote a book on the descent of man on the Moon and wrote another book on a submarine, the Nautilus, which made the world tour in immersion.

But these fantastic tales were not based on any scientific or technological content; a man was shot on the moon like a bullet and the Nautilus was certainly not powered by atomic energy.

But here, for the first time, we are faced with a law that defines the evolution of a process over time and that has proved true for many years and on the basis of which we can project the same process for future years.

We are able to do something different from that of a science fiction writer, although wonderful. We own an exponential law whose mathematical extension to the future can tell us where we're going to end.

To prevent this exercise from becoming a trivial mathematical calculation that even a boy of the middle schools would be able to calculate, we must first make some background clarification.

Let's resume the concept underlying the digital world, that is, all the systems, including the computers, use as a fundamental cell, an element that has only two states: zero and one. We might think of building any digital system with switches which can be of any shape, and not just Transistors.

In fact, our integrated circuits are nothing more than a set of elements, precisely the Transistors, which can take only the two states, zero and one, and whose integration has proven to be fantastically achievable industrially.

Attention, this is the key point of the argument we're going to specify: these Transistors are actually nothing but switches that can take the state on or off, they have nothing special, apart from the way of how they are made as we have seen.

What does this mean? That theoretically we could build any digital system starting from an element that can be built in order to present two states, open or closed: it can be a valve, a faucet, a switch or whatever comes to mind.

Of course it would not be easy to build a computer using taps and water instead of electric current, but conceptually this is possible ... it would be rather large and a quite slow computer, but definitely possible.

Why we do this reasoning? Because this is the needed premise if we want to take some consideration on the extension of Moore's law to a far distant future, avoiding fiction ruminations and then to be able to make some realistic assumption.

Now, scientists' prediction is that, in addition to a certain size, i.e. when the Transistor becomes too small, its operation is no longer possible.

And this is true: if we approach too much the size of an atom it is clear that the Transistor, as we mean it today, no longer can exist, i.e. it can no longer operate as Transistor, and therefore that dimension is the actual theoretical limit that will be reached in less than ten years.

We could however go on to three dimensions, instead of the current two, i.e. making integrated circuits in cubes rather than mere platelets, but even so it would stretch Moore's law just a few years, maybe to get production until 2025 or 2027 but not much beyond.

But then what kind of prediction can we reasonably expect for the year 2050? Here's where we have to get back to the basics of digital, that is the basic unit that opens and closes by generating zero and one.

If we extend the Moore's law to elements other than the Transistor, but much smaller and able to perform the fundamental function of a switch, we can foresee already from today extremely complex systems far more complex than the today ones.

Here we are in the area of the possible, not yet of the certain, but at least of the probable. Who can today deny that one day the technicians will not be able to use as a switch a molecule or even an atom?

For some time we have been talking about nanotechnologies, as an example, a sector that has not yet entered the common jargon, but that scientists have been studying for years.

It is not here the case of going into the details of these technologies, but we cite them just to indicate how investigations are proceeding far beyond the limits predictable by Moore's law applied to the Transistors.

More, the hypothesis of extending Moore's law until the year 2050 is not so farfetched, and we shouldn't care if it will be realized through the qubits of modern quantum computers or a Transistor as big as an atom or as large as a quark or some other devilry that can assume one or zero status.

We must not forget that the electronics industry has long based its existence on this kind of growth and therefore fearful investments for these advances are certainly a visible reality and that gives us hope.

If the research will continue, in the year 2050 we will be able to buy an electronic device that will contain microchips with at least a quadrillion Transistors, namely a thousand of tera Transistors, or ten to the power of 15.

Note that these gigantic numbers are already in the popular jargon today and came to us with mass storage devices.

As I write these lines, I came with an exceptional offer by a major retailer for a hard disk with a capacity of three terabytes and who knows how much memory I'll be able to buy at the same price in the year 2050? Surely at least some million of terabytes.

But what can be done with so much processing power and memory? And here's the catch: it is certain that we cannot foresee and answer to such a question, but I like to come back to what I wrote in 1969 at the end of my first book on integrated circuits and that sound so:

"... as always happens in the technique, it is not excluded that this technology, once perfected and made economic, leads to the realization of devices suitable for applications in other sectors, for now not yet foreseeable."

And then the only sector that was considered for the digital chips was that of computers and certainly none, not even science fiction writers, could predict that in our days a two-year-old child with a few hundred million Transistors in his hand could choose the favorite cartoons, resident in a server thousands of miles away, and that he could choose with the simple touch of a finger!!

The only thing we can do now is proposing a bold comparison between the complexity of a digital system and the complexity of the brain of living beings from a microbe to a man.

Microchip progress in comparison with the development of living beings

It must be clear that complexity does not necessarily mean reaching the superior functions of the human brain, as we already said, but this confrontation allows us to affirm that the miniaturization of digital circuits will surely exceed the one created by nature and that thanks to this human ingenuity we will surely be able to go further and to create unimaginable tools of progress and hopefully useful for all.

In fact, if we believe that the synapses are the smallest element on which the brain builds its elaborations and knowing that we have about one million billion of them, that is, ten elevated to the fifteenth power, and if we consider a synapse equivalent to a Transistor, then our number of synapses will be crossed by a microchip right in the year 2050.

If Moore's law will hold up to the year 2050 at that date the industry will be able to produce a single microchip containing many Transistors, or many components similar to a switch, as many as are the synapses contained in a human brain.

I repeat, this does not mean that we will have the exact equivalent of our brain in electronic form: synapses and what nature has given us in four billion years of evolution will remain substantially different, but this comparison outline what kind of opportunities our semiconductor technology, in a few decades, will offer us to enjoy.

The Role of Other Nations

Now we know how the Transistor became the hero of Silicon Valley and, not only of that Valley, because soon all the industrialized nations were flung in pursuit of American industry, not least Italy with two major companies born in the late fifties namely Società Generale Semiconduttori (SGS and now STMicroelectronics) and Elsi-Raytheon, then turned ATES and today part of STMicroelectronics.

In Europe the Dutch Philips was dominant and from which, while student, I bought my first diodes and Transistors, the French were in this race too with Thomson-Houston and the Germans with Siemens and Intermetall.

Japan understood the importance of the Transistors and, as we have seen, began in the 50s to invest in products manufactured for the consumer market, with Sony in the front row. The Japanese companies dominated the consumer electronics and perhaps only now they can be undermined by countries such as China and South Korea.

It must be said that the US industry was very engaged in military and space applications while Japan could devote all its resources to the civil sector and this allowed them to maintain an advantage that it would prove unattainable even by the best American companies.

And so, while Americans were descending on the Moon, the Europeans were competing with the Americans, the Russians were competing with both, the Japanese "swallowed" all the consumer electronics industry of the West with their walkmans, HI FI stereos, TV sets and cameras.

But the superb Silicon Valley has struck back and some of those companies that had taken root there shortly after the man had reached the Moon have created new immense markets and are today's leaders with their innovative products.

I am referring to Intel with 80% of the world's CPU market, to Apple Computer with the unreachable iphone, iPad and the rest invented by the ingenious Steve Jobs, to Microsoft that, had further north, dominates the market of operating systems.

On a my recent trip I have been able to see how much is being developed even in non-traditional areas for Silicon Valley such as photovoltaic energy, biology and medicine.

Silicon Valley

In this book I could not miss to talk about the territory where everything we have so far spoken about has been developed: namely the Silicon Valley.

After the birth in 1956 of the first silicon company, founded by William Shockley, the Shockley Semiconductor Laboratory, in that valley the silicon crystal became the basic element to produce solid state electronic components such as diodes, Transistors, integrated circuits and microprocessors.

Its purification and its use to fabricate these various chips became what gold had been for job seekers of the wild west two centuries before!

In restaurants, cafés, even in the parks there was no person who did not speak of semiconductors, silicon, integrated circuits.

I remember phrases like: "We are producing excellent MSI (read Medium Scale Integration)" or "We are already ahead with LSI (read Large Scale Integration) and soon we will pass to the VLSI (Very Large Scale Integration)" and other diabolical phrases that were part of a jargon now disappeared.

I believe that even the children there were playing at who was better with silicon or at games that involved the most popular topic at the time, like making money at the monopoly of integrated circuits.

The Silicon Valley owes everything to that Transistor of 1947 from which it flourished one of the most advanced industries in the world and that has made the United States the spearhead of the electronic sector and then the computer industry: they should make a monument to the Transistor and probably a monument exists somewhere!

Here is that a journalist in 1971 decides to call that area south of San Francisco, between Palo Alto and San Josè and flourishing of companies producing all sorts of silicon chips as "Silicon Valley", an expression which has since spread worldwide when one want to define a production area with the presence of high-tech companies.

As mentioned several times, I was fortunate enough to frequent the Silicon Valley in those years and I took advantage of those trips to collect extensive photographic documentation and to have many interesting meetings, information and contacts that I still keep with great care.

I still have the Rich's Business Guide of Silicon Valley dated 1984 that displays thousands of companies in the electronics industry, their categories and the number of employees whose total calculated for all companies listed in the Guide reaches hundreds of thousands.

This 200 pages guide was my Bible when visiting the Silicon Valley

I like report here the note that appears at the first page of the Guide and that represents well the situation of then and perhaps even today:

"Santa Clara County's Silicon Valley has the greatest concentration of Hi Tech companies in the world. The nature of this industry creates constant change in not only the "State of the Art" but also in company location, configuration, and products of the technical "artists" and entrepreneurs.

Therefore, overnight, small companies emerge into giants, companies merge, companies go public and many people become instant millionaires, people leave a company and spawn several small companies hoping to find the magic product or service; some succeed while others fail and again seek new monies or ideas.

This exciting innovative climate, that recognizes with respect both success and failure, is in great part responsible for the United States' continual leadership in the Hi Tech world.

This constant change makes it difficult for those who do business in Silicon Valley to keep informed of what company is producing what product and how they can be contacted. Rich's Business Guide provides the most comprehensive, up to date information on the Electronics industry, answering those very pertinent questions."

The Rich's Guide defines more exactly the Silicon Valley as the territory extending from South San Josè, including the Santa Clara County and going north to Mountain View, near Palo Alto and the world famous Stanford University.

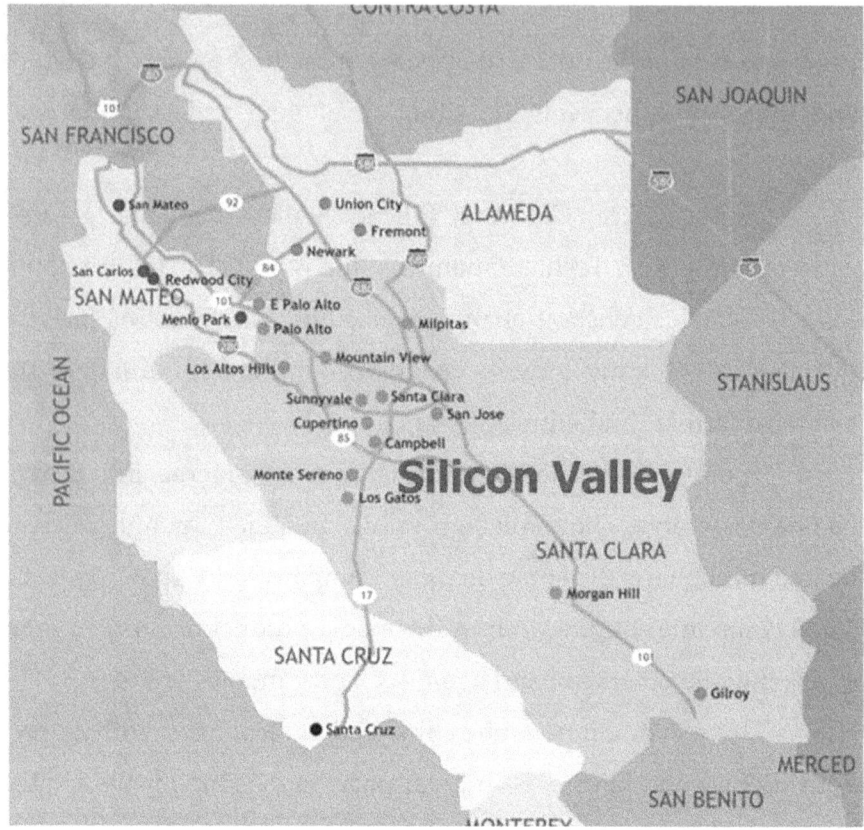

Silicon Valley took its name in the early 1970s. Extending south of Palo Alto to Santa Clara and surroundings

To reach the Silicon Valley today, as then, landing at San Francisco International Airport, from there you take the 101 Freeway South and crossing Red Wood City, Carmel, you move to Palo Alto, Sunnyvale and then, if traffic permits, in three quarters of an hour you can take one of the four exits to Santa Clara.

There you penetrate the area of the highest concentration of companies, including Intel, Yahoo, Google and many others, and a little further South, at no more than ten minutes, there is Cupertino with the immense headquarters for Apple Computer, while moving North it develops the Google building, which in my days did not exist and indeed the area was forested with a dense vegetation, now disappeared.

In August 2010 I passed back to that area and with some nostalgia I have reviewed those places and some of those companies that I had met and with whom I had treated many years before.

In front of the parking lot of the Apple, in Cupertino, I saw and photographed the old motorcycle of Steve Jobs, sign that he was inside the establishment, but I did not dare to enter to greet him and to remember our first meetings at the beginning of the eighties in the office of Mike Markkula, the then CEO of the company, occasion in which I was presented in preview the Apple III.

I regret I didn't get into the headquarters and I didn't see Steve for the last time.

Steve Jobs' motorcycle parked in front of his factory

I then visited the headquarters of Intel in Santa Clara, a company that I've seen the birth of and with whom I have worked for almost twenty years as an independent rep.

So much time had passed since my first encounter when Intel resided in Mountain View at Middlefield Rd and now I read on the face of this enormous and recent construction "This building is dedicated to Robert Noyce", who died prematurely in 1990 and I had the honor and pleasure to meet and attend in the early 1970s for so many times.

Noyce founded Intel together with Dr. Moore in August 1968 and I remember well when in 1970 I accompanied him to visit Olivetti at Ivrea and Honeywell at Pregnana Milanese, driving the car rather quickly on the highway Milan-Turin and, like with other Intel officers, I was asked slow down.

I remember even when some time later, and always on the same highway, accompanying to Ivrea Gordon Moore, Ed Gelbach and Tom Lawrence and trying to arrive at the appointment on time by running at over 100 miles/hour at some point Ed yelled from behind: "If you do not slow down immediately, you will lose the franchisee ... crazy Ettore, you have half of Intel in this car! ".

So I learned again how important was to respect speed limits in the US (I risked to be arrested in California on freeway 101 in 1971 for that!).

Coming back to the lovely Silicon Valley, I would like here to express my opinion of how such an incredible concentration of companies, talents and research fervor could happen and why I feel it difficult, if not impossible, to reproduce such a situation in other parts of the world.

I'm sure there have been four conditions, hardly repeatable: first I put definitely the beautiful environment, almost constant temperature, never too hot, never too cold, a Valley not far from a Pacific Ocean that keeps the temperature always fresh and still isolated from the ocean by a range of hills that weakens the wind and atmospheric turbulence, so ideal for going to live as well as working.

Second the timing: I mean that there has been a magical period in our recent history in which in the United States, and not only there, there was a fervor to do, to grow, to develop, fervor that was born from being released from the frightening second world war with the desire to forget and to build a new world.

Third, I put the presence of an important university with technological tendencies and entrepreneurial spirit; I refer to Stanford University.

I know many people put this university at first place, but I don't agree. In the United States there are other universities of similar rank, but I think that the natural environment that encircled the university caused anyone to stay there to live and this has certainly played a fundamental role. We all like to work, earn good money, but if the place where we work is also one of the best in the world, why to go elsewhere.

If an important engineer was leaving his company for any reason, it was there that he tried to find another job or to found his new business and was not seeking a job in another State. The Italian Faggin's case is a good example.

Fourth, I refer to the name "Silicon Valley" and its ready available venture capitals that made the valley the world Capital for that . which defines the modern capital of technology.

The name of the valley owes it to William Shockley, the 1956 Nobel Prize for the invention of the Transistor, that in the same year left the Bell Laboratories, on the Atlantic coast, and went to Mountain view to found its Shockley semiconductor laboratory.

From which company then, in 1958, the traitorous eight (as Shockley defined them) founded Fairchild Semiconductor from which , in turn, came out Noyce and Moore to found Intel, Jerry Sanders to found AMD, and others to found Intersil, Nitron, General Microelectronics, Amelco, Signetics, Integrated Electronics, Avenced Memory Systems etc., etc, and so on in cascade for many years.

As a distant observer and having spoken to many locals, I can conclude, as I believe anywhere, has played a basic role what I said, that people lived well there and had no interest to move elsewhere and the valley has been always plentiful of both jobs and entrepreneurship.

By the way I have to add that now the times have changed and not because the place has become less welcoming, but for the costs soared to the stars for housing, for the land and the labor and many companies have moved to other States much less expensive like Oregon and Texas, but the big heads of companies hold headquarters in Silicon Valley and guess why.

Conclusion

That Semimetallic Monsterling born in 1947 is the progenitor of everything we've described here, is the progenitor of incredible technological achievements that have made possible the realization of many of reverie that in the television series Star Trek were offered to the public!

I remember the little video communicator with which Captain Kirk communicated with his spaceship, very similar to a today smartphone with which hundreds of millions people communicate with each other, only that the smartphone is clearly much superior than the communicator of Captain Kirk

The thin panel-shaped monitors and TV sets present in billions of houses, the robots controlled by voice and appliances that we can control almost as living beings.

In many cases we have exceeded all expectations and who knows that one day, as in science fiction, we could achieve teleportation as in the Star-Trek series. Anyway, in the medical field, with the application of highly advanced instrumentation, we have overcome many of the wonders we could imagine.

More, who could have thought that thanks to a mountain of Transistors strung in a chip ultra complex we would be able to remote control a robot that walks freely and alone on the planet Mars while on the Earth a group of scientists command it and it walks, observes, drills, collects, analyzes and tells us what's on Mars.

> **This is nothing compared to what awaits us in the future, be sure!**

Author's blog: http://ettoreaccenti.blogspot.com